Additional and Drill Manual

Cathy Ferrer
Valencia Community College

College Algebra and Trigonometry
Fourth Edition

and

Precalculus
Fourth Edition

Margaret L. Lial
American River College

John Hornsby
University of New Orleans

David I. Schneider
University of Maryland

PEARSON
Addison Wesley

Boston San Francisco New York
London Toronto Sydney Tokyo Singapore Madrid
Mexico City Munich Paris Cape Town Hong Kong Montreal

Reproduced by Pearson Addison-Wesley from electronic files supplied by the author.

Copyright © 2009 Pearson Education, Inc.
Publishing as Pearson Addison-Wesley, 75 Arlington Street, Boston, MA 02116.

All rights reserved. No part of this publication may be reproduced, stored in a retrieval system, or transmitted, in any form or by any means, electronic, mechanical, photocopying, recording, or otherwise, without the prior written permission of the publisher. Printed in the United States of America.

ISBN-13: 978-0-321-52928-2
ISBN-10: 0-321-52928-6

1 2 3 4 5 6 BB 10 09 08 07

Contents

Chapter R **Review of Basic Concepts**

R.1 Sets 1
R.2 Real Numbers and Their Properties 2
R.3 Polynomials 3
R.4 Factoring Polynomials 4
R.5 Rational Expressions 5
R.6 Rational Exponents 6
R.7 Radical Expressions 7
Answers 8

Chapter 1 **Equations and Inequalities**

1.1 Linear Equations 12
1.2 Applications and Modeling with Linear Equations 13
1.3 Complex Numbers 14
1.4 Quadratic Equations 15
1.5 Applications and Modeling with Quadratic Equations 16
1.6 Other Types of Equations and Applications 17
1.7 Inequalities 18
1.8 Absolute Value Equations and Inequalities 19
Answers 20

Chapter 2 **Graphs and Functions**

2.1 Rectangular Coordinates and Graphs 24
2.2 Circles 25
2.3 Functions 26
2.4 Linear Functions 27
2.5 Equations of Lines; Curve Fitting 28
2.6 Graphs of Basic Functions 29
2.7 Graphing Techniques 30
2.8 Function Operations and Composition 31
Answers 32

Chapter 3 **Polynomial and Rational Functions**

3.1 Quadratic Functions and Models 37
3.2 Synthetic Division 38
3.3 Zeros of Polynomial Functions 39
3.4 Polynomial Functions: Graphs, Applications, and Models 40
3.5 Rational Functions: Graphs, Applications, and Models 41
3.6 Variation 42
Answers 43

Chapter 4   **Inverse, Exponential, and Logarithmic Functions**

    4.1 Inverse Functions   47
    4.2 Exponential Functions   48
    4.3 Logarithmic Functions   49
    4.4 Evaluating Logarithms and the Change-of-Base Theorem   50
    4.5 Exponential and Logarithmic Equations   51
    4.6 Applications and Models of Exponential Growth and Decay   52
    Answers   53

Chapter 5   **Trigonometric Functions**

    5.1 Angles   56
    5.2 Trigonometric Functions   57
    5.3 Evaluating Trigonometric Functions   58
    5.4 Solving Right Triangles   59
    Answers   60

Chapter 6   **The Circular Functions and Their Graphs**

    6.1 Radian Measure   63
    6.2 The Unit Circle and Circular Functions   64
    6.3 Graphs of the Sine and Cosine Functions   65
    6.4 Translations of the Graphs of the Sine and Cosine Functions   66
    6.5 Graphs of the Tangent, Cotangent, Secant, and Cosecant Functions   67
    6.6 Harmonic Motion   68
    Answers   69

Chapter 7   **Trigonometric Identities and Equations**

    7.1 Fundamental Identities   73
    7.2 Verifying Trigonometric Identities   74
    7.3 Sum and Difference Identities   75
    7.4 Double-Angle and Half-Angle Identities   76
    7.5 Inverse Circular Functions   77
    7.6 Trigonometric Equations   78
    7.7 Equations Involving Inverse Trigonometric Functions   79
    Answers   80

Chapter 8   **Applications of Trigonometry**

    8.1 The Law of Sines   84
    8.2 The Law of Cosines   85
    8.3 Vectors, Operations, and the Dot Product   86
    8.4 Applications of Vectors   87
    8.5 Trigonometric Form of Complex Numbers; Products and Quotients   88
    8.6 DeMoivre's Theorem; Powers and Roots of Complex Numbers   89
    8.7 Polar Equations and Graphs   90
    8.8 Parametric Equations, Graphs, and Applications   91
    Answers   92

Chapter 9   **Systems and Matrices**

    9.1 Systems of Linear Equations   97
    9.2 Matrix Solution of Linear Systems   98
    9.3 Determinant Solution of Linear Systems   99
    9.4 Partial Fractions   100
    9.5 Nonlinear Systems of Equations   101
    9.6 Systems of Inequalities and Linear Programming   102
    9.7 Properties of Matrices   103
    9.8 Matrix Inverses   104
    Answers   105

Chapter 10   **Analytic Geometry**

    10.1 Parabolas   110
    10.2 Ellipses   111
    10.3 Hyperbolas   112
    10.4 Summary of the Conic Sections   113
    Answers   114

Chapter 11   **Further Topics in Algebra**

    11.1 Sequences and Series   117
    11.2 Arithmetic Sequences and Series   118
    11.3 Geometric Sequences and Series   119
    11.4 The Binomial Theorem   120
    11.5 Mathematical Induction   121
    11.6 Counting Theory   122
    11.7 Basics of Probability   123
    Answers   124

Appendix A   **Polar Form of Conic Sections**   127
Appendix B   **Rotation of Axis**   128
    Answers   129

R.1 Sets

Use set notation, and list all the elements of each set.

1. $\{20, 21, 22, ..., 29\}$

2. $\{4, 8, 12, ..., 28\}$

3. $\{\frac{1}{3}, \frac{1}{9}, \frac{1}{27}, ..., \frac{1}{729}\}$

4. $\{51, 48, 45, ... 39\}$

Identify the following sets as finite or infinite.

5. $\{9, 18, 27, ...\}$

6. $\{1, \frac{1}{3}, \frac{1}{9}, ...\}$

7. $\{x \mid x \text{ is a natural number less than } 6\}$

8. $\{x \mid x \text{ is a rational number}\}$

Complete the blanks with either $\in$ or $\notin$ so that the resulting statement is true.

9. 9 _____ $\{3, 6, 9, 12, 15, 18\}$

10. $\emptyset$ _____ $\{\emptyset\}$

11. -2 _____ $\{x \mid x \text{ is a natural number} < 6\}$

12. $\{4\}$ _____ $\{0, 2, 4, 8, 16\}$

Tell whether each statement is true or false.

13. $19 \notin \{14, 16, 18, ...\}$

14. $\{5, 19, 45, 4\} = \{4, 5, 9, 45\}$

15. $\{2, 4, 6\} \cap \{1, 3, 5\} = \emptyset$

16. $A \cup \emptyset = A$

Let $U = \{1, 2, 3, 4, ..., 20\}$, $A = \{2, 4, 6, ... 20\}$, $B = \{1, 5, 9, 13, 17\}$, $C = \{3, 6, 9, 12, 15, 18\}$. **Tell whether each statement is true or false.**

17. $C \subseteq A$

18. $\emptyset \subseteq C$

19. $B \subseteq U$

20. $A \subseteq B$

Let $U = \{1, 2, 3, 4, ..., 10\}$, $A = \{2, 4, 6, ... 10\}$, $B = \{1, 3, 5, ..., 9\}$, $C = \{3, 6, 9\}$. **Use these sets to find each of the following.**

21. $A \cap B$

22. $A \cup B$

23. $B \cap C$

24. $A \cup C$

R.2 Real Numbers and Their Properties

Let set $A = \{-3, \frac{2}{3}, \frac{7}{0}, \pi, \sqrt{13}, -\frac{8}{2}, \frac{1}{9}, 5\}$. **List all the elements of A that belong to each set.**

1. Natural numbers
2. Integers
3. Rational numbers
4. Irrational numbers

Evaluate each expression.

5. $(-2)^3$
6. -3^2
7. $-3 \cdot 2^4$
8. 5^3

Evaluate each expression.

9. $3^2 - 8 \div 4$
10. $(9+3) \div 4 \cdot 2 + 6$
11. $5 - 2(4^2 - 7)$
12. $\dfrac{-(-2)^5 - (3-5)}{3(2) + 4(-6)}$

13. $(5^2 - \sqrt{81})(-2^3 + 8)$
14. $36 \div 6 \cdot 2 \div 4 + 4$
15. $3^3 - 2(5 - 6 \cdot 3)$
16. $\left(\dfrac{2}{3} + \dfrac{4}{5}\right) - \left(\dfrac{1}{6} - \dfrac{1}{3}\right)$

Evaluate each expression if $x = -1$, $y = 2$ and $z = -3$.

17. $2x^2 + 8y - z$
18. $\dfrac{x+3}{4} - \dfrac{z^3}{y}$
19. $2z \div y \cdot x$
20. $3x^2 + 3xy + z^2$

Simplify each expression:

21. $11 - (x - 3)$
22. $\dfrac{2xh + h^2}{h}$
23. $\dfrac{2}{3}\left(\dfrac{15}{4}x - \dfrac{27}{14}y + \dfrac{9}{2}\right)$

Evaluate each expression.

24. $|5|$
25. $\left|-\dfrac{9}{3}\right|$
26. $-\left|\dfrac{3}{4}\right|$
27. $-\left|\dfrac{31}{3}\right|$

Let $x = 3$ and $y = -4$. Evaluate each expression.

28. $|x + y|$
29. $|x| + |y|$
30. $\dfrac{2|x| + 2|y|}{|xy|}$
31. $\dfrac{|3x - 2y|}{-|y|}$

Simplify each expression.

1. $2^3 \, 2^2$

2. $(-3x^3)(2x^6)$

3. $(8y^2)(4y^{10})$

4. $(-2x^2y^3)^3$

5. $\left(\dfrac{x^5}{y^{10}}\right)^2$

Identify each expression as a polynomial or not a polynomial. For each polynomial, give the degree and identify it as a monomial, binomial, trinomial, or none of these.

6. $m^3 + 2m$

7. $x^2 + 4x + 4$

8. $6x^2y^3 + 11x^4y^2$

9. $\sqrt{2}m^3 - \sqrt{5}m^2 + 4m + 6$

10. $\dfrac{6}{x^2} + \dfrac{8}{x} + \dfrac{2}{3}$

Find each sum or difference.

11. $(3x^2 - 4x + 8) + (4x^2 + 5x - 1)$

12. $(2y^3 + 5y + 9) + (3y^2 - 2y + 5)$

13. $(4x^2 + 8x^5 - 7) - (4x^2 - 3x - 6)$

14. $(2x^2 + 4x - 1) - (3x^2 - 2x - 4) + (8x^2 + 11)$

Find each product.

15. $(2x + 3)(3x - 4)$

16. $(x - \tfrac{1}{2})(x + \tfrac{3}{4})$

17. $(-2y^2)(3y^3 - 6y^2 + 5y - 7)$

18. $(x + 3)(5x^4 + 2x^2 - 6x - 1)$

Find each product.

19. $(3x + 4)(3x - 4)$

20. $(2y + 3)^3$

21. $(3x - 6)^2$

22. $[(2x + 3) + 6][(2x + 3) - 6]$

Perform the indicated operations.

23. $2(2x^2 + 3x - 6) - 3(x^2 - 6x - 1)$

24. $x^3 \cdot 2x - (1 + x^2) \cdot 3x^2$

25. $-x^2(7x - 3) - 4x(2x + 3)$

26. $(4x^3 - 6x^2)(x - 1) - (x^4 - 2x^3)$

R.4 Factoring Polynomials

Factor out the greatest common factor from each polynomial.

1. $3x^4 - 18x^3 - 12x^2$

2. $10x^2 - 24x^4 + 12x^6$

3. $x^3 + 3x^2h + 3xh^2 + h^3 - x^3$

4. $(x-1)^3 + 4(x-1)^2 + 3(x-1)$

Factor each polynomial by grouping.

5. $xy - x + 4y - 4$

6. $6xy + 14x - 15y - 35$

7. $2y + 10 - 3xy - 15x$

8. $x^3 + 3x^2 + 4x + 12$

Factor each trinomial.

9. $2x^2y + 16xy + 30y$

10. $3x^2 + 5x + 2$

11. $5x^2 - 18x - 8$

12. $24x^4y + 30x^3y^2 - 9x^2y^3$

Factor each polynomial.

13. $9x^2 - 81y^2$

14. $25x^2 - 16$

15. $x^3 - 8$

16. $x^4 + 64x$

17. $(x+1)^3 - 64$

Factor each polynomial by substitution.

18. $3m^4 + 14m^2 + 8$

19. $10(x-1)^2 + (x-1) - 2$

20. $15(2-x)^2 + 2(2-x) - 24$

21. $3x^6 - 7x^3 - 40$

R.5 Rational Expressions

Find the domain of each rational expression.

1. $\dfrac{2x+1}{x-7}$
2. $\dfrac{x-2}{x^2+4x+3}$
3. $\dfrac{4}{x^2+2x+4}$

Write each rational expression in lowest terms.

4. $\dfrac{3x-9}{4x-12}$
5. $\dfrac{2x-4}{4x-8}$
6. $\dfrac{3x-6}{x^2-7x+10}$

7. $\dfrac{x^2-25x}{5x}$
8. $\dfrac{2x^2+x-3}{2x^2+7x+6}$

Find each product or quotient.

9. $\dfrac{8x^2}{24y} \div \dfrac{3x}{9y^2}$
10. $\dfrac{4x+6}{12} \cdot \dfrac{6}{2x+3}$

11. $\dfrac{x^2+xy}{20x-25} \cdot \dfrac{4x-5}{x^2}$
12. $\dfrac{x^2+2x-15}{x^2-9} \div \dfrac{2x^2+11x+5}{2x^2+7x+3}$

13. $\dfrac{x^2+x-2}{x^2+3x-4} \cdot \dfrac{x^2+2x-8}{x^2-3x-10}$
14. $\dfrac{4x^2-21x+5}{2x^2-x-1} \cdot \dfrac{3-2x-x^2}{x^2+2x+1} \div \dfrac{16x^2-1}{1-x^2}$

Perform each addition or subtraction.

15. $\dfrac{1}{4x}+\dfrac{2x}{7}$
16. $\dfrac{5x^2}{7}-\dfrac{7}{10x}$
17. $\dfrac{x^2+x}{x}+\dfrac{x+1}{2x-5}$

18. $\dfrac{x+1}{x}-\dfrac{x-5}{x+3}$
19. $\dfrac{x(3x-1)}{x^2+1}+\dfrac{x-1}{x+1}$
20. $\dfrac{1}{3x}+\dfrac{1}{5(2x-1)}-\dfrac{1}{15(x+2)}$

Simplify each expression.

21. $\dfrac{\dfrac{1}{x}+2}{4-\dfrac{1}{x}}$
22. $\dfrac{3+\dfrac{1}{x+4}}{\dfrac{1}{x-4}-1}$

R.6 Rational Exponents

Write each expression with only positive exponents and evaluate if possible. Assume all variables represent nonzero real numbers.

1. -2^{-3}
2. 7^{-3}
3. $3x^{-2}$
4. $-6y^{-2}$

Perform the indicated operations. Write each answer using only positive exponents. Assume all variables represent nonzero real numbers.

5. $\dfrac{x^2}{x^8}$
6. $\dfrac{y^{-3}}{y^{-4}}$
7. $\dfrac{16x^{-2}y^{-3}}{4x^{-3}y^5}$
8. $\dfrac{24x^3(y^3)^{-2}}{8x^4y^{-7}}$

Simplify each expression.

9. $81^{1/4}$
10. $16^{1/2}$
11. $\left(\dfrac{-8}{27}\right)^{1/3}$
12. $(-81)^{1/2}$

Perform the indicated operations. Write each answer using only positive exponents. Assume all variables represent positive real numbers.

13. $16^{3/2}$
14. $\dfrac{25^{-7/2}}{25^{-3}}$
15. $-\dfrac{2}{3}x^{-2/3} \cdot x^{-1}$
16. $\dfrac{(y^{2/3})^4}{(y^{-1})^{1/6}}$

17. $\left(\dfrac{x^4y^6}{16}\right)^{1/2}\left(\dfrac{8x^{-3}}{y^9}\right)^{1/3}$
18. $\dfrac{x^{4/7}x^{1/14}x^{1/2}}{(x^9)^{-1/3}}$

Find each product. Assume that all variables represent positive real numbers.

19. $x^{3/2}(2x^{4/3} - 6x^{1/7})$
20. $\dfrac{1}{3}x^{-2/3}(x+x^2)$

21. $(x^{1/2} - x^{1/3})(x + x^{1/3})$
22. $(y^{1/3} + y^{-1/3})^2$

Factor, using the given common factor. Assume all variables represent positive real numbers.

23. $x^{-16} - x^{-8}$, x^{-16}
24. $y^{8/9} + 27y^{-7/9}$, $y^{-7/9}$

25. $(x+1)^{-4/3} + (x+1)^{-1/3} + (x+1)^{1/3}$, $(x+1)^{-4/3}$
26. $(2x-1)^{-7/5} + (2x-1)^{1/5} + (2x-1)^{3/5}$, $(2x-1)^{-7/5}$

R.7 Radical Expressions

Write in radical form. Assume all variables represent positive real numbers.

1. $(6x-4y)^{2/3}$
2. $(5+2y)^{1/7}$

Write in exponential form. Assume all variables represent positive real numbers.

3. $\sqrt[7]{x^3}$
4. $-y\sqrt[4]{5x^9}$

Simplify each expression. Assume all variables represent positive real numbers.

5. $\sqrt{(2x-1)^2}$
6. $-\sqrt[4]{625x^{13}y^9}$
7. $-\sqrt[3]{\dfrac{81}{125}}$
8. $\sqrt{\dfrac{x^7 y^{10}}{z^{11}}}$
9. $-\sqrt[3]{\dfrac{16x^5}{y^{13}z^8}}$
10. $\sqrt[4]{\sqrt{x}}$

Simplify each expression, assume all variables represent positive numbers.

11. $6\sqrt{18}-3\sqrt{2}+9\sqrt{8}$
12. $\sqrt[3]{3}+5\sqrt[3]{24}-9\sqrt[3]{81}$
13. $(\sqrt{6}+3)(\sqrt{6}-3)$
14. $(\sqrt{x}+\sqrt[3]{x})^2$
15. $\dfrac{\sqrt[4]{16x^5 y^3}\,\sqrt[4]{x^3 y^6}}{\sqrt[4]{81x^4 y}}$
16. $\dfrac{9x\sqrt{x}-x}{\sqrt{x}}$

Rationalize the denominator of each radical expression. Assume all variables represent nonnegative numbers and that no denominators are 0.

17. $\dfrac{\sqrt{2}}{\sqrt{3}-\sqrt{2}}$
18. $\dfrac{1+\sqrt{6}}{\sqrt{6}+\sqrt{3}}$
19. $\dfrac{\sqrt{y}}{7-\sqrt{y}}$
20. $\dfrac{5x}{2-\sqrt{x+y}}$

Answers

R.1 Sets

1. {20, 21, 22, 23, 24, 25, 26, 27, 28, 29}
2. {4, 8, 12, 16, 20, 24, 28}
3. $\left\{\dfrac{1}{3}, \dfrac{1}{9}, \dfrac{1}{27}, \dfrac{1}{81}, \dfrac{1}{243}, \dfrac{1}{729}\right\}$
4. {51, 48, 45, 42, 39}
5. infinite
6. infinite
7. finite
8. infinite
9. $\in$
10. $\in$
11. $\notin$
12. $\notin$
13. true
14. false
15. true
16. true
17. false
18. true
19. true
20. false
21. $\varnothing$
22. U
23. C
24. {2,3,4,6,8,9,10}

R.2 Real Numbers and Their Properties

1. 5
2. $-3, -\dfrac{8}{2}, 5$
3. $-3, \dfrac{2}{3}, -\dfrac{8}{2}, \dfrac{1}{9}, 5$
4. $\pi, \sqrt{13}$
5. -8
6. -9
7. -48
8. 125
9. 7
10. 12
11. -13
12. $-\dfrac{17}{9}$
13. 0
14. 7
15. 53
16. $\dfrac{49}{30}$
17. 21
18. 14
19. 3
20. 6
21. $14 - x$
22. $2x + h$
23. $\dfrac{5}{2}x - \dfrac{9}{7}y + 3$
24. 5
25. 3
26. $-\dfrac{3}{4}$
27. $-\dfrac{31}{3}$
28. 1
29. 7
30. $\dfrac{7}{6}$
31. $\dfrac{17}{4}$

Answers

R.3 Polynomials

1. 32

2. $-6x^9$

3. $32y^{12}$

4. $-8x^6y^9$

5. $\dfrac{x^{10}}{y^{20}}$

6. polynomial, 3, binomial

7. polynomial, 2, trinomial

8. polynomial, 6, binomial

9. polynomial, 3, none of these

10. not a polynomial

11. $7x^2 + x + 7$

12. $2y^3 + 3y^2 + 3y + 14$

13. $8x^5 + 3x - 1$

14. $7x^2 + 6x + 14$

15. $6x^2 + x - 12$

16. $x^2 + \dfrac{1}{4}x - \dfrac{3}{8}$

17. $-6y^5 + 12y^4 - 10y^3 + 14y^2$

18. $5x^5 + 15x^4 + 2x^3 - 19x - 3$

19. $9x^2 - 16$

20. $8y^3 + 36y^2 + 54y + 27$

21. $9x^2 - 36x + 36$

22. $4x^2 + 12x - 27$

23. $x^2 + 24x - 9$

24. $-x^4 - 3x^2$

25. $-7x^3 - 5x^2 - 12x$

26. $3x^4 - 8x^3 + 6x^2$

R.4 Factoring Polynomials

1. $3x^2(x^2 - 6x - 4)$

2. $2x^2(5 - 12x^2 + 6x^4)$

3. $h(3x^2 + 3xh + h^2)$

4. $x(x-1)(x+2)$

5. $(x+4)(y-1)$

6. $(2x-5)(3y+7)$

7. $-(3x-2)(y+5)$

8. $(x+3)(x^2+4)$

9. $2y(x+3)(x+5)$

10. $(x+1)(3x+2)$

11. $(x-4)(5x+2)$

12. $3x^2y(2x+3y)(4x-y)$

13. $9(x-3y)(x+3y)$

14. $(5x-4)(5x+4)$

15. $(x-2)(x^2+2x+4)$

16. $x(x+4)(x^2-4x+16)$

17. $(x-3)(x^2+6x+21)$

18. $(m^2+4)(3m^2+2)$

19. $(2x-1)(5x-7)$

20. $(3x-10)(5x-4)$

21. $(x^3-5)(3x^3+8)$

Answers

R.5 Rational Expressions

1. $\{x \mid x \neq 7\}$ **2.** $\{x \mid x \neq -3, -1\}$ **3.** $(-\infty, \infty)$ **4.** $\dfrac{3}{4}$

5. $\dfrac{1}{2}$ **6.** $\dfrac{3}{x-5}$ **7.** $\dfrac{x-25}{5}$ **8.** $\dfrac{x-1}{x+2}$

9. xy **10.** 1 **11.** $\dfrac{x+y}{5x}$ **12.** 1

13. $\dfrac{x-2}{x-5}$ **14.** $\dfrac{(x-5)(x+3)(x-1)}{(2x+1)(x+1)(4x+1)}$ **15.** $\dfrac{8x^2+7}{28x}$ **16.** $\dfrac{50x^3-49}{70x}$

17. $\dfrac{2x^2-2x-4}{2x-5}$ **18.** $\dfrac{9x+3}{x^2+3x}$ **19.** $\dfrac{4x^3+x^2-1}{x^3+x^2+x+1}$ **20.** $\dfrac{11x^2+22x-10}{30x^3+45x^2-30x}$

21. $\dfrac{1+2x}{4x-1}$ **22.** $\dfrac{3x^2+x-52}{-x^2+x+20}$ **23.** $\dfrac{x^2+x}{1-x}$ **24.** $\dfrac{x^3-x-1}{2x}$

R.6 Rational Exponents

1. $-\dfrac{1}{8}$ **2.** $\dfrac{1}{343}$ **3.** $\dfrac{3}{x^2}$ **4.** $\dfrac{-6}{y^2}$

5. $\dfrac{1}{x^6}$ **6.** y **7.** $\dfrac{4x}{y^8}$ **8.** $\dfrac{3y}{x}$

9. 3 **10.** 4 **11.** $-\dfrac{2}{3}$ **12.** not a real number

13. 64 **14.** $\dfrac{1}{5}$ **15.** $-\dfrac{2}{3x^{5/3}}$ **16.** $y^{17/6}$

17. $\dfrac{x}{2}$ **18.** $x^{29/7}$ **19.** $2x^{17/6} - 6x^{23/14}$ **20.** $\dfrac{1}{3}x^{1/3} + \dfrac{1}{3}x^{4/3}$

21. $x^{3/2} + x^{5/6} - x^{4/3} - x^{2/3}$ **22.** $y^{2/3} + 2 + \dfrac{1}{y^{2/3}}$ **23.** $x^{-16}(1-x^8)$ **24.** $y^{-7/9}(y^{5/3} + 27)$

25. $(x+1)^{-4/3}\left(1 + (x+1) + (x+1)^{5/3}\right)$ **26.** $(2x-1)^{-7/5}\left(1 + (2x-1)^{8/5} + (2x-1)^2\right)$

R.7 Radical Expressions

1. $\sqrt[3]{(6x-4y)^2}$ **2.** $\sqrt[7]{5+2y}$ **3.** $x^{3/7}$ **4.** $-5^{1/4}x^{9/4}y$

5. $|2x-1|$ **6.** $-5x^3y^2\sqrt[4]{xy}$ **7.** $-\dfrac{3}{5}\sqrt[3]{3}$ **8.** $\dfrac{x^3y^5}{z^5}\sqrt{\dfrac{x}{z}}$

9. $-\dfrac{2x}{y^4z^2}\sqrt[3]{\dfrac{2x^2}{yz^2}}$ **10.** $x^{1/8}$ **11.** $33\sqrt{2}$ **12.** $-16\sqrt[3]{3}$

13. -3 **14.** $x + 2\sqrt[6]{x^5} + \sqrt[3]{x^2}$ **15.** $\dfrac{2xy^2}{3}$ **16.** $9x - \sqrt{x}$

17. $\sqrt{6} + 2$ **18.** $\dfrac{\sqrt{6} - \sqrt{3} + 6 - 3\sqrt{2}}{3}$ **19.** $\dfrac{7\sqrt{y} + y}{49 - y}$ **20.** $\dfrac{10x + 5x\sqrt{x+y}}{4 - x - y}$

1.1 Linear Equations

Solve each equation.

1. $4x - 9 = 11x - 2$

2. $\dfrac{7}{8}x + 6 - 2x = \dfrac{3}{4}$

3. $3[x - (6 - 3x) + 5] = 11x + 8$

4. $0.5 - 0.8x = 1.4x + 0.6$

5. $\dfrac{4}{5}x - 0.6x + 0.3 = x - 9$

6. $5 - (x + 3) = 3x + 4(x - 9)$

7. $\dfrac{1}{5}(x - 3) = \dfrac{1}{2}(x - 1) - \dfrac{2}{3}x$

8. $3(8 + 6x) = 12$

9. $-0.2 - (0.4x - 0.3) = -1.8 - 0.6x$

Decide whether each equation is an identity, a conditional equation, or a contradiction. Give the solution set.

10. $6(4x - 1) + 3 = 2(4x - 1 + 8x) - 1$

11. $-0.4x + 0.75 + 2(0.8 + 0.2x) = -0.6$

12. $3x - 2 + 4(8x + 9) = 2(6 + x)$

Solve each formula for the indicated variable. Assume that the denominator is not 0 if variables appear in the denominator.

13. $v^2 = v_0^2 + \dfrac{2Fd}{m}$ for F

14. $v = \pi r^2 h$ for h

15. $A = P + Prt$ for r

16. $\beta = -\dfrac{1}{V} \cdot \dfrac{dV}{dP}$ for V

Solve each equation for x.

17. $5(3a + x) = ab - c$

18. $\dfrac{a}{x + 4} = ab + c$

19. $a - 4bx = (a + x)(a + 3y)$

20. $a^2 - ax + b^2 = bx - c$

In the metric system of weights and measures, temperature is measured in degrees Celsius ($°C$) instead of degrees Fahrenheit ($°F$). To convert back and forth between the two systems, we use the equations $C = \dfrac{5}{9}(F - 32)$ and $F = \dfrac{9}{5}C + 32$. In each exercise, convert to the other system. Round answers to the nearest tenth of a degree if necessary.

21. $0°C$

22. $74°F$

1.2 Applications and Modeling with Linear Equations

Solve each problem.

1. The length of a calculus book is 2 inches more than the width. The perimeter is 36 inches. Find the width.

2. An avid gardener wants to make a circular flowerbed with a circumference of 32 feet. What should the radius be?

3. The perimeter of a rectangle is 82 feet. Find the length and width of the rectangle if the length is 7 feet more than the width.

4. The perimeter of an isosceles triangle is 28 inches. Each of the two equal sides measures 9 inches. What is the length of the third side?

5. Scott drives to work in the morning using the toll roads and he averages 60 mph. In the afternoon he drives home on free roads and he averages 45 mph. It took him 50 minutes longer to make the trip in the afternoon. How far does Scott live from work?

6. Rachel paddles her canoe upstream on the Econ River for 45 minutes. Coming downstream with the same boat speed took 30 minutes. If the current in this river is 3 mph, what was her boat speed?

7. Suppose that tuition costs $69.69 per credit hour and that student fees are fixed at $25.00. Antonio paid $1070.35 for his fall classes. How many credit hours did he sign up for?

8. How many gallons of a 9% peroxide solution must be mixed with 3 gallons of a 15% peroxide solution to obtain a 12% solution?

9. How much pure acid should be added to 16 ml of a 21% acid solution to increase the concentration to 30%?

10. Mr. Davis has $7000 to invest in two accounts. Part of the money will be invested at 3% and the rest will be invested at 4%. Find the amount invested at each rate if the total annual income from interest is $240.

11. Ms. Gordon wants to invest $9000 in stocks. She invests part of the money in a stock that pays 4% annual interest. The rest is invested in a business that pays 5% interest. Find the amount invested in each account if the total annual income from interest is $430.

12. Dan is 3 times as old as Mo, and in 6 years he will be twice as old. How old are Dan and Mo now?

13. A mother is 24 years older than her daughter. In 18 years, the mother's age will be twice that of her daughter. How old are they now?

1.3 Complex Numbers

Identify each number as real, complex, or pure imaginary.

1. 3
2. $4i$
3. $2-7i$

Write each number as the product of a real number and *i*.

4. $\sqrt{-16}$
5. $\sqrt{-7}$
6. $-\sqrt{-36}$
7. $-\sqrt{-120}$

Multiply or divide as indicated. Simplify each answer.

8. $\sqrt{-9}\sqrt{16}$
9. $\sqrt{-2}\sqrt{-32}$
10. $\dfrac{\sqrt{-48}}{\sqrt{12}}$
11. $\dfrac{-\sqrt{30}}{\sqrt{-90}}$
12. $\dfrac{\sqrt{-8}\sqrt{-4}}{\sqrt{2}}$

Write each number in standard form $a+bi$.

13. $\dfrac{-8+\sqrt{-48}}{4}$
14. $\dfrac{18-\sqrt{-72}}{6}$

Find each sum or difference. Write the answer in standard form.

15. $(4+5i)+(2+3i)$
16. $(7-6i)+(2-3i)$
17. $(2+5i)-(3+4i)$
18. $(4-3i)-(3-4i)$

Find each product. Write the answer in standard form.

19. $(5-2i)(3+4i)$
20. $(2+5i)^2$
21. $i(4-3i)^2$
22. $(1+i)(1-i)(2+3i)(2-3i)$

Simplify each power of *i*.

23. i^{28}
24. i^{29}
25. i^{-109}

Find each quotient. Write the answer in standard form.

26. $\dfrac{2+5i}{4+i}$
27. $\dfrac{-3+2i}{-1-4i}$
28. $\dfrac{-7}{-i}$

1.4 Quadratic Equations

Solve each equation by the zero-factor property.

1. $x^2 + 2x = 8$
2. $3x^2 - 11x = -6$

Solve each equation by the square root property.

3. $x^2 = -36$
4. $(3x-1)^2 = -12$

Solve each equation by completing the square.

5. $x^2 + 2x + 4 = 0$
6. $x^2 + 14x + 24 = 0$
7. $18x^2 - 9x = 20$
8. $2x^2 - 6x + 1 = 0$

Solve each equation using the quadratic formula.

9. $x^2 - 6x - 27 = 0$
10. $5x^2 - 6x = 9$
11. $5x^2 + 6x = -9$
12. $x^2 + 7x - 3 = 0$
13. $x^2 + 2x - 4 = 0$

Solve each cubic equation.

14. $x^3 - 64 = 0$
15. $x^3 + 125 = 0$
16. $8x^3 = 27$

Solve each equation for the indicated variable. Assume no denominators are 0.

17. $F = \dfrac{km_1 m_2}{d^2}$ for d
18. $A = P\left(1 + \dfrac{r}{n}\right)^2$ for r

Evaluate the discriminant for each equation. Then use it to predict the number of distinct solutions, and whether they are rational, irrational or nonreal complex. Do not solve the equation.

19. $x^2 - 5x - 14 = 0$
20. $2x^2 - 7x - 8 = 0$
21. $3x^2 - 2x + 1 = 0$

For each pair of numbers, find the values of a, b, and c for which the quadratic equation $ax^2 + bx + c = 0$ has the given numbers as solutions.

22. $-6, 2$
23. $2 + \sqrt{3},\ 2 - \sqrt{3}$
24. $4i,\ -4i$

1.5 Applications and Modeling with Quadratic Equations

Solve each problem.

1. A farmer wants to set up a pigpen using 40 feet of fence to enclose a rectangular area of 51 square feet. Find the dimensions of the pigpen.

2. A vegetable gardener has a tilled piece of soil that measures 6 yards by 12 yards. She wants to put a border of marigolds of the same width within the four sides and leave an area of 40 square yards for vegetables. How wide can the border be?

3. A playground has an area of 456 meters and a perimeter of 100 meters. How long and wide is it?

4. A classroom has an area of 952 square feet. The length of the room is 6 feet more than its width. How long and wide is it?

5. Frank is decorating a porch that measures 10 yards by 20 yards. He wants to paint a uniform border around the sides and leave 96 square yards in the middle of the floor for a faux rug painting. How wide should he paint the border?

6. A rectangular piece of cardboard is 5 inches longer than it is wide. Three-inch squares are cut out of each corner and the resulting flaps are turned up and taped to form an open box. If the volume of the resulting box is 18 square inches, what were the original dimensions of the piece of cardboard?

7. A can of tuna has a surface area of 48.69 square inches. Its height is 1.87 inches. What is the radius of the can?

8. The volume of an 18-ounce box of cereal is 4524 cubic centimeters. The width of the box is 11.5 cm less than the length and the height is 29 cm. Find the length and width of the box.

9. The volume of a box of Jell-O is 9.7 cubic inches. The width of the box is 0.5 inches less than the length and its depth is 1 inch. Find the length and width of the box to the nearest thousandth.

10. A fish is being pulled onto a boat with a line attached to the fish at water level. When the fish is 10 feet from the boat, the length of the fishing line from the boat to the fish is 2 feet longer than the height of the fishing pole from the water. Find the height of the fishing pole.

11. A kite is flying on 20 feet of string. Its horizontal distance from the person flying it is 4 feet less than its vertical distance from the ground. Find its horizontal distance from the person and its vertical distance form the ground.

12. The size of a rectangular TV screen is given by the length of its diagonal. The length of the screen is 23 inches more than its width. Find the dimensions of a 63-inch TV screen.

1.6 Other Types of Equations and Applications

Solve each equation.

1. $\dfrac{3}{x+1} = \dfrac{x}{x+1} + 2$

2. $\dfrac{6}{x-3} - \dfrac{x+2}{x-3} = 4$

3. $\dfrac{5}{x} + 6 = \dfrac{1}{x+2}$

4. $\dfrac{6}{x^2+x-2} + \dfrac{3}{x^2-x} = \dfrac{5}{x-1}$

5. $\dfrac{x}{x-4} - 2 = \dfrac{8}{x-4}$

6. $\dfrac{4}{x-2} - \dfrac{1}{2} = \dfrac{3x-8}{2x-4}$

7. $\dfrac{1}{x+1} = \dfrac{1}{3} - \dfrac{1}{6x+6}$

8. $\dfrac{4}{3x} + \dfrac{3}{3x+1} = -2$

9. $\dfrac{x+3}{x+4} + \dfrac{x+4}{x+3} = \dfrac{2x^2-16}{x^2+7x+12}$

10. $\dfrac{4x-8}{3x^{2/3}} = 0$

Solve each equation.

11. $\sqrt{x-3} = x - 5$

12. $\sqrt{x-2} - \sqrt{x} = 7$

13. $\sqrt{4x^2+4} = 2x+1$

14. $\sqrt{2x+9} - \sqrt{x+1} = \sqrt{x+5}$

15. $\sqrt{x+4} + \sqrt{2x-1} = 3\sqrt{x-1}$

16. $5 + \sqrt{x-6} = 3$

17. $1 + \sqrt[3]{x+3} = 7$

18. $\sqrt[4]{3x+4} = \sqrt[4]{x-9}$

19. $\sqrt{17x - \sqrt{x^2-5}} = 7$

20. $x^{5/3}(6-x)^{4/3} = 0$

Solve each equation for the indicated variable. Assume all denominators are nonzero.

21. $A = 2x\sqrt{r^2-x^2}$ for r

22. $\dfrac{1}{R} = \dfrac{1}{A} + \dfrac{1}{B}$ for B

23. $F = \dfrac{Gm_1 m_2}{d^2}$ for d

24. $I = \dfrac{E}{r+R}$ for r

1.7 Inequalities

Solve each inequality. Write each solution set in interval notation.

1. $4x-1 < x+8$

2. $3x+4 < x+10$

3. $9(x+2)-4 \leq 10-x$

4. $\dfrac{1-2x}{3}+\dfrac{x}{2} \leq x-\dfrac{1+3x}{6}$

5. $6(x+2) > 2(2x+5)$

Solve each inequality. Write each solution set in interval notation.

6. $-11 \leq 3(x-1)-2 \leq 16$

7. $1 < \dfrac{3-2x}{5} \leq \dfrac{7}{3}$

8. $4 < x+2 \leq 16$

9. $1 < \dfrac{4x-3}{2} \leq 8$

10. $-2 \leq \dfrac{10-x}{2} \leq 6$

Solve each quadratic inequality. Write each solution set in interval notation.

11. $x^2 - 5x + 6 \leq 0$

12. $x^2 - x - 12 < 0$

13. $x^2 < 100$

14. $x^2 + 4x > 5$

15. $36x^2 - 24x \leq 24$

Solve each rational inequality. Write each solution set in interval notation.

16. $3x - 2 < \dfrac{10-x}{2}$

17. $\dfrac{x+3}{2} \geq \dfrac{2-x}{-2}$

18. $\dfrac{4x-3}{2} \geq \dfrac{2-x}{3}$

19. $\dfrac{1}{x-6} \leq 3$

20. $\dfrac{x+4}{x-2} \geq 2$

1.8 Absolute Value Equations and Inequalities

Solve each equation.

1. $|x-3|=6$

2. $|x-1|=9$

3. $\left|\dfrac{4x-7}{3}\right|=11$

4. $|9-4x|=|7+2x|$

5. $|3x-4|=|4-3x|$

Solve each inequality. Give the solution set using interval notation.

6. $|x-3|\le 1$

7. $|x-1|<3$

8. $2|x-6|>9$

9. $\left|\dfrac{1}{2}x-\dfrac{1}{3}\right|<4$

10. $|7x-9|\ge 15$

11. $|9x-2|-6\le 12$

12. $|2x+8|-9\ge 2$

13. $\left|\dfrac{15x-1}{5}\right|\le 3$

14. $|3x-4|-4\ge 9$

15. $|4x+3|-11\le 15$

Solve each equation or inequality.

16. $|7x+2|\ge 0$

17. $|9x-4|\ge -2$

18. $|4x+3|=0$

19. $|5x-2|\le 0$

20. $|6x+1|\le -3$

20 Answers

1.1 Linear Equations

1. $\{-1\}$
2. $\{\frac{14}{3}\}$
3. $\{11\}$
4. $\{-0.045\}$
5. $\{11.625\}$
6. $\{\frac{19}{4}\}$
7. $\{\frac{3}{11}\}$
8. $\{-\frac{2}{3}\}$
9. $\{-9.5\}$
10. identity
11. contradiction
12. conditional, $\{-\frac{2}{3}\}$
13. $F = \dfrac{m(v^2 - v_0^2)}{2d}$
14. $h = \dfrac{v}{\pi r^2}$
15. $r = \dfrac{A - P}{Pt}$
16. $V = -\dfrac{1}{\beta} \cdot \dfrac{dV}{dP}$
17. $x = \dfrac{ab - c - 15a}{5}$
18. $x = \dfrac{a}{ab+c} - 4$
19. $x = \dfrac{a^2 + 3ay - a}{-4b - a - 3y}$
20. $x = \dfrac{a^2 + b^2 + c}{a+b}$
21. $32°F$
22. $23.3°C$

1.2 Applications and Modeling with Linear Equations

1. 8 inches
2. 5.1 feet
3. 17 feet, 24 feet
4. 10 inches
5. 150 miles
6. 15mph
7. 15 credits
8. 3 gallons
9. 2.1 ml
10. $4000 at 3%; $3000 at 4%
11. $2000 at 4%; $7000 at 5%
12. Don is 18; Mo is 6
13. Mother is 30; daughter is 6

1.3 Complex Numbers

1. real, complex
2. pure imaginary
3. complex
4. $4i$
5. $\sqrt{7}i$
6. $-6i$
7. $-2\sqrt{30}i$
8. $12i$
9. -8
10. $2i$
11. $\dfrac{\sqrt{3}}{3}i$
12. -4
13. $-2 + \sqrt{3}i$
14. $3 - \sqrt{2}i$
15. $6 + 8i$
16. $9 - 9i$

17. $-1+i$
18. $1+i$
19. $23+14i$
20. $-21+20i$

21. $24+7i$
22. 26
23. 1
24. i

25. $-i$
26. $\frac{13}{17}+\frac{18}{17}i$
27. $-\frac{5}{17}-\frac{14}{17}i$
28. $-7i$

1.4 Quadratic Equations

1. $\{-4,2\}$
2. $\{\frac{2}{3},3\}$
3. $\{-6i,6i\}$
4. $\{\frac{1\pm 2\sqrt{3}i}{3}\}$

5. $\{-1\pm\sqrt{3}i\}$
6. $\{-12,-2\}$
7. $\{-\frac{5}{6},\frac{4}{3}\}$
8. $\{\frac{3\pm\sqrt{7}}{2}\}$

9. $\{-3,9\}$
10. $\{\frac{3\pm 3\sqrt{6}}{5}\}$
11. $\{-\frac{3}{5}\pm\frac{6}{5}i\}$
12. $\{\frac{-7\pm\sqrt{61}}{2}\}$

13. $\{-1\pm\sqrt{5}\}$
14. $\{4,-2\pm 2\sqrt{3}i\}$
15. $\{-5,\frac{5}{2}\pm\frac{5\sqrt{3}}{2}i\}$
16. $\{\frac{3}{2},-\frac{3}{4}\pm\frac{3\sqrt{3}}{4}i\}$

17. $d=\pm\sqrt{\frac{k m_1 m_2}{F}}$
18. $r=-n\pm n\sqrt{\frac{A}{P}}$
19. 2 distinct rational solutions
20. 2 distinct irrational solutions

21. 2 distinct nonreal complex solutions
22. $a=1, b=4, c=-12$
23. $a=1, b=-4, c=1$
24. $a=1, b=0, c=16$

1.5 Applications and Modeling with Quadratic Equations

1. 17 feet by 3 feet
2. 1 yard
3. 12 m by 38 m
4. 28 feet by 34 feet

5. 2 yards
6. 7 in. by 12 in.
7. 2 inches
8. 19.5 cm by 8 cm

9. 2.875 in. by 3.375 in.
10. 24 feet
11. 12 feet by 16 feet
12. 31.5 in. by 54.5 in.

1.6 Other Types of Equations and Applications

1. $\{\frac{1}{3}\}$
2. $\{\frac{16}{5}\}$
3. $\{-\frac{5}{3},-1\}$
4. $\{-\frac{6}{5}\}$
5. $\{0\}$
6. $\{\frac{9}{2}\}$
7. $\{\frac{5}{2}\}$
8. $\{-\frac{1}{6},-\frac{4}{3}\}$
9. $\{-\frac{41}{14}\}$
10. $\{2\}$
11. $\{7\}$
12. $\varnothing$
13. $\{\frac{3}{4}\}$
14. $\{-\frac{1}{2}\}$
15. $\{5\}$
16. $\varnothing$
17. $\{213\}$
18. $\{-\frac{13}{2}\}$
19. $\{3\}$
20. $\{0,6\}$
21. $r = \pm\sqrt{\frac{A^2}{4x^2}+x^2}$
22. $B = \frac{AR}{A-R}$
23. $d = \pm\sqrt{\frac{G\,m_1\,m_2}{F}}$
24. $r = \frac{E-IR}{I}$

1.7 Inequalities

1. $(-\infty,3)$
2. $(-\infty,3)$
3. $(-\infty,-\frac{2}{5}]$
4. $[\frac{3}{4},\infty)$
5. $(-1,\infty)$
6. $[-2,7]$
7. $[-\frac{13}{3},-1)$
8. $(2,14]$
9. $(\frac{5}{4},\frac{19}{4}]$
10. $[-2,14]$
11. $[2,3]$
12. $(-3,4)$
13. $(-10,10)$
14. $(-\infty,-5)\cup(1,\infty)$
15. $[\frac{1-\sqrt{7}}{3},\frac{1+\sqrt{7}}{3}]$
16. $(-\infty,2)$
17. $(-\infty,\infty)$
18. $[\frac{13}{14},\infty)$
19. $(-\infty,6]\cup[\frac{19}{3},\infty)$
20. $[2,8]$

1.8 Absolute Value Equations and Inequalities

1. $\{-3, 9\}$
2. $\{-8, 10\}$
3. $\{-\frac{13}{2}, 10\}$
4. $\{\frac{1}{3}, 8\}$

5. $(-\infty, \infty)$
6. $[2, 4]$
7. $(-2, 4)$
8. $\left(-\infty, \frac{3}{2}\right) \cup \left(\frac{21}{2}, \infty\right)$

9. $\left(-\frac{22}{3}, \frac{26}{3}\right)$
10. $(-\infty, -\frac{6}{7}] \cup [\frac{24}{7}, \infty)$
11. $[-\frac{16}{9}, \frac{20}{9}]$
12. $(-\infty, -\frac{19}{2}] \cup [\frac{3}{2}, \infty)$

13. $[-\frac{14}{15}, \frac{16}{15}]$
14. $(-\infty, -3] \cup \left[\frac{17}{3}, \infty\right)$
15. $[-\frac{29}{4}, \frac{23}{4}]$
16. $(-\infty, \infty)$

17. $(-\infty, \infty)$
18. $\{-\frac{3}{4}\}$
19. $\{\frac{2}{5}\}$
20. $\varnothing$

2.1 Rectangular Coordinates and Graphs

For the points P and Q, find the distance $d(P,Q)$ and the coordinates of the midpoint of the segment PQ.

1. $(1,4), (-6,-1)$
2. $(-2,-5), (1,7)$
3. $(0,9), (2,19)$
4. $(-5,12), (1,-11)$

Determine whether the three points are the vertices of a right triangle.

5. $(-2,1), (2,3), (2,1)$
6. $(-3,3), (2,-2), (3,3)$
7. $(10,40), (30,-20), (10,-20)$
8. $(-75,-50), (75,-25), (25,25)$

Determine whether the three points are collinear.

9. $(-2,3), (0,3), (2,3)$
10. $(-2,0), (0,-2), (3,-2)$
11. $(1,2), (2,1), (-1,2)$
12. $(-3,-2), (0,0), (3,2)$

Find the coordinates of the other endpoint of each segment, given its midpoint and one endpoint.

13. midpoint $(-9,-3)$, endpoint $(-5,-7)$
14. midpoint $(-1,-1)$, endpoint $(2,-5)$
15. midpoint $(5,2)$, endpoint $(4,6)$
16. midpoint $(-4.5,4)$, endpoint $(-2,13)$

For each equation, give three ordered pairs that are solutions.

17. $x + y = 7$
18. $4x = 4y + 16$
19. $y = x^2 - 6$
20. $y = -x^2$
21. $y = x^3 + 3$
22. $y = 2 - x^3$
23. $y = -|x - 7|$
24. $y = \sqrt{x} + 5$

Find the center-radius form of the equation of each circle.

1. center $(0,0)$; radius 5

2. center $(0,0)$; radius 1

3. center $(0,0)$; radius 3

4. center $(0,0)$; radius $\sqrt{7}$

5. center $(1,2)$; radius 1

6. center $(-3,-1)$; radius 2

7. center $(1,-3)$; radius 7

8. center $(-2,3)$; radius $\sqrt{5}$

9. center $(3,4)$; radius 3

10. center $(-5,-7)$; radius $\sqrt{2}$

Decide whether or not each equation has a circle as its graph. If it does, give the center and the radius.

11. $x^2 - 6x + y^2 + 2y - 15 = 0$

12. $x^2 - 2x + y^2 - 3 = 0$

13. $4x^2 + 9y^2 - 16x + 18y - 11 = 0$

14. $x^2 + 4x + y^2 - 6y = 3$

15. $x^2 + y^2 = 16$

16. $x^2 + y^2 - 12y = -32$

17. $x^2 + y^2 - 8x = -10$

18. $x^2 + y^2 + 2x - 6y = -6$

19. $x^2 + y^2 + 8x + 12y = -54$

20. $x^2 + y^2 - 4x - 6y = -13$

2.3 Functions

Decide whether each relation defines a function and give the domain and range.

1. $\{(-4,0),(-2,1),(0,2),(3,9)\}$
2. $\{(-2,7),(2,3),(-2,0),(5,-2)\}$
3. $\{(3.5,2.5),(2.0,2.0),(2.5,4.0),(3.5,3.5)\}$
4. $\{(50,10),(100,13),(150,14),(250,13)\}$

Decide whether each relation defines y as a function of x. Give the domain and range.

5. $y = x^6$
6. $y = \sqrt{4-2x}$
7. $2x - 2y < 8$
8. $xy = 2$

Let $f(x) = x^2 - 6x$ **and** $g(x) = 2x+1$. **Find the following.**

9. $f(-3)$
10. $f(a+h)$
11. $g\left(-\dfrac{1}{2}\right)$
12. $g(3t-4)$

For each function, find $f(-1)$ and $f(3)$.

13. $f = \{\left(-1,\dfrac{13}{3}\right),(0,4),(3,3),(12,0)\}$
14. $f(x) = 3x^4 - 4x^3$
15. $f(x) = 1.75x$
16. $f(x) = 3 - |x|$

An equation that defines y as a function of x is given. Solve for y in terms of x and represent y with the function notation $f(x)$.

17. $3x - 4y = 5$
18. $y - 2x^2 = 4$
19. $x + 3y = 6$
20. $y + 6x^2 = 2$

Determine the intervals of the domain for which each function is (a) increasing, (b) decreasing, (c) constant.

21. $f(x) = |x| - 2$
22. $f(x) = \sqrt{x} - 3$
23. $f(x) = (x+1)^3$
24. $f(x) = (x-3)^2 + 4$

Graph each linear function. Identify any constant functions. Give the domain and the range.

1. $f(x) = x + 2$

2. $f(x) = -x - 1$

3. $f(x) = 4$

4. $f(x) = \dfrac{2}{3}x + \dfrac{5}{3}$

5. $f(x) = 3x$

Graph each vertical line. Give the domain and range of the relation.

6. $x = 2$

7. $x + 3 = 0$

8. $2x - 6 = 0$

Find the slope of the line satisfying the given conditions.

9. through $(-2, -8)$ and $(1, 1)$

10. through $(-3, 5)$ and $(1, 5)$

11. through $\left(\dfrac{2}{3}, -\dfrac{5}{2}\right)$ and $\left(\dfrac{1}{2}, -\dfrac{5}{3}\right)$

12. horizontal, through $(3, -4)$

13. vertical, through $(-1, 2)$

Find the slope of the line.

14. $y = -2x + 3$

15. $y = x - 1$

16. $2y - 4x = 8$

17. $4y + 3x = 12$

18. $-2y = 6x$

19. $y + 1 = 2(x - 6)$

20. $2(y - 8) = 6(x + 3)$

2.5 Equations of Lines; Curve Fitting

Write an equation for the line described. Give answers in standard form.

1. through $(1,3)$; $m = \dfrac{2}{3}$

2. through $(0,0)$; $m = 0$

3. through $(7,-2)$ and $(-3,4)$

4. through $(6,1)$; undefined slope

Write an equation for the line described. Give answers in slope-intercept form (if possible).

5. through $(-3,-2)$; $m = 6$

6. through $\left(\dfrac{7}{8}, \dfrac{3}{4}\right)$ and $\left(\dfrac{4}{5}, \dfrac{1}{2}\right)$

7. vertical; through $(2,7)$

8. horizontal; through $(-1,6)$

Give the slope and y-intercept of each line.

9. $x + 2y = 5$

10. $y = \dfrac{1}{2}x + 4$

11. $4y - x = -12$

12. $3x - 2y = 4$

Write an equation in standard form and in slope-intercept form for the line described.

13. through $(1,2)$; parallel to $y - 2x = 3$

14. through $(2,-7)$; perpendicular to $y = -\dfrac{5}{3}x - 2$

15. through $(3,-4)$; parallel to $y + 5x = 7$

16. through $(-10,3)$; perpendicular to $y = 3$

Determine whether the three points are collinear by using slopes.

17. $(-2,5), (0,3), (2,1)$

18. $(2,4), (4,8), (8,16)$

19. $(-1,3), (-4,0), (-7,-3)$

20. $(6,2), (1,3), (5,7)$

2.6 Graphs of Basic Functions

Determine the intervals of the domain over which each function is continuous.

1. $f(x) = \begin{cases} x+2 & \text{for } x \leq -1 \\ \frac{1}{2}x+2 & \text{for } x > -1 \end{cases}$

2. $f(x) = \begin{cases} \frac{5}{3}x+2 & \text{for } x \leq 0 \\ 3-x & \text{for } x > 0 \end{cases}$

3. $f(x) = \begin{cases} 3x-12 & \text{for } x < 3 \\ \frac{1}{2}x+3 & \text{for } x \geq 3 \end{cases}$

4. $f(x) = \begin{cases} 3-\frac{1}{2}x & \text{for } x \leq -2 \\ 2x+8 & \text{for } x > -2 \end{cases}$

For each piecewise defined function, find (a) $f(-4)$, **(b)** $f(-2)$, **(c)** $f(0)$, **and (d)** $f(2)$.

5. $f(x) = \begin{cases} x+2 & \text{for } x \leq -1 \\ \frac{1}{2}x+2 & \text{for } x > -1 \end{cases}$

6. $f(x) = \begin{cases} \frac{5}{3}x+2 & \text{for } x \leq -6 \\ 3-x & \text{for } x > -6 \end{cases}$

7. $f(x) = \begin{cases} \frac{3}{4}x & \text{for } x < -2 \\ -\frac{1}{2}x+5 & \text{for } -2 \leq x < 1 \\ 6-3x & \text{for } x \geq 1 \end{cases}$

8. $f(x) = \begin{cases} 7-x & \text{for } x < -3 \\ 3 & \text{for } -3 \leq x \leq 2 \\ -2x+2 & \text{for } x > 2 \end{cases}$

9. $f(x) = \begin{cases} 3+x & \text{for } x < -2 \\ x & \text{for } -2 \leq x \leq 0 \\ -2x & \text{for } x > 0 \end{cases}$

10. $f(x) = \begin{cases} 4x+2 & \text{for } x < -3 \\ \frac{1}{4}x-2 & \text{for } -3 \leq x \leq 1 \\ -x & \text{for } x > 1 \end{cases}$

11. $f(x) = \begin{cases} 3x-12 & \text{for } x \leq 0 \\ \frac{1}{2}x+3 & \text{for } x > 0 \end{cases}$

12. $f(x) = \begin{cases} 3-\frac{1}{2}x & \text{for } x \leq -2 \\ 2x+8 & \text{for } x > -2 \end{cases}$

13. $f(x) = \begin{cases} 5-3x & \text{for } x < 1 \\ 4+3x & \text{for } x \geq 1 \end{cases}$

14. $f(x) = \begin{cases} -4x-2 & \text{for } x < -1 \\ x & \text{for } x \geq -1 \end{cases}$

15. $f(x) = \begin{cases} 3x-1 & \text{for } x < -2 \\ 3x+1 & \text{for } -2 \leq x \leq 2 \\ -\frac{1}{3}x-1 & \text{for } x > 2 \end{cases}$

16. $f(x) = \begin{cases} -4x & \text{for } x \leq -1 \\ 3 & \text{for } -1 < x < 1 \\ 2x & \text{for } x \geq 1 \end{cases}$

2.7 Graphing Techniques

Given the function $f(x)$, describe how to obtain the graph of $g(x)$.

1. $f(x) = x^2$; $g(x) = -3x^2$

2. $f(x) = x^2$; $g(x) = \dfrac{1}{3}x^2$

3. $f(x) = x^2$; $g(x) = (-3x)^2$

4. $f(x) = |x|$; $g(x) = 2|x|$

5. $f(x) = |x|$; $g(x) = -2|x|$

6. $f(x) = \sqrt{x}$; $g(x) = \dfrac{1}{2}\sqrt{x}$

Without graphing, determine whether each equation has a graph that is symmetric with respect to the x-axis, the y-axis, the origin, or none of these.

7. $y = 2x^2 + 3$

8. $y = -3x^4$

9. $y^2 = 4 - x^2$

10. $y = 2x^3$

11. $y = 2x + 1$

12. $y = \sqrt{x^2 - 4}$

Decide whether each function is even, odd, or neither.

13. $y = x^2 + 2$

14. $y = 3x^3 - 1$

15. $y = 0.25x^4 - x^2 + 1$

16. $y = x^5 - x^4$

17. $y = x^2 + |x|$

18. $y = |x| - x$

Given the function $f(x)$, describe how to obtain the graph of $g(x)$.

19. $f(x) = x^2$; $g(x) = x^2 + 1$

20. $f(x) = x^2$; $g(x) = x^2 - 3$

21. $f(x) = x^2$; $g(x) = (x+1)^2$

22. $f(x) = \sqrt{x}$; $g(x) = \sqrt{x-3}$

23. $f(x) = |x|$; $g(x) = |x-1| + 2$

24. $f(x) = x^3$; $g(x) = (x-2)^3$

2.8 Function Operations and Composition

Let $f(x) = 4x^2 + 2x$ **and let** $g(x) = 5x - 4$. **Find each of the following.**

1. $(f+g)(2)$
2. $(fg)(2)$
3. $\left(\dfrac{f}{g}\right)(1)$

For the pair of functions defined, find $f+g$, $f-g$, fg, **and** $\dfrac{f}{g}$. **Give the domain of each.**

4. $f(x) = 6x + 4$
   $g(x) = x - 3$

5. $f(x) = x^2 - 1$
   $g(x) = 2x^2 - x + 1$

6. $f(x) = \sqrt{2x+1}$
   $g(x) = \dfrac{1}{x}$

Use the table to evaluate each expression, if possible, (a) $(f+g)(1)$, **(b)** $(fg)(-2)$, **(c)** $\left(\dfrac{f}{g}\right)(0)$.

7.

x	f(x)	g(x)
-2	4	2
-1	2	2
0	5	1
1	-4	6

8.

x	f(x)	g(x)
-2	4	1
-1	1	1
0	-1	-3
1	2	2

9.

x	f(x)	g(x)
-2	-1	0
-1	1	1
0	-2	0
1	-1	2

For each of the functions defined as follows, find (a) $f(x+h)$, **(b)** $f(x+h) - f(x)$, **and (c)** $\dfrac{f(x+h) - f(x)}{h}$.

10. $f(x) = 2x - 1$
11. $f(x) = x^3$
12. $f(x) = x^2 - 2$

Let $f(x) = 3x - 2$ **and** $g(x) = -2x - 1$. **Find each function value.**

13. $(f \circ g)(-1)$
14. $(g \circ f)(3)$
15. $(f \circ f)(4)$

Find $(f \circ g)(x)$ **and** $(g \circ f)(x)$ **for each pair of functions.**

16. $f(x) = 4x + 2$
    $g(x) = 6 - 3x$

17. $f(x) = 2x + 2$
    $g(x) = 3x^2 - x + 2$

18. $f(x) = 5x^2 + 4x$
    $g(x) = \sqrt{x - 1}$

Answers

2.1 Rectangular Coordinates and Graphs

1. $d = \sqrt{74}$; $\left(-\frac{5}{2}, \frac{3}{2}\right)$ **2.** $d = 3\sqrt{17}$; $\left(-\frac{1}{2}, 1\right)$ **3.** $d = 2\sqrt{26}$; $(1, 14)$ **4.** $d = \sqrt{565}$; $\left(-2, \frac{1}{2}\right)$

5. yes **6.** no **7.** yes **8.** no

9. yes **10.** no **11.** no **12.** yes

13. $(-13, 1)$ **14.** $(-4, 3)$ **15.** $(6, -2)$ **16.** $(-7, -5)$

17. Answers will vary. **18.** Answers will vary. **19.** Answers will vary. **20.** Answers will vary.

21. Answers will vary. **22.** Answers will vary. **23.** Answers will vary. **24.** Answers will vary.

2.2 Circles

1. $x^2 + y^2 = 25$ **2.** $x^2 + y^2 = 1$ **3.** $x^2 + y^2 = 9$ **4.** $x^2 + y^2 = 7$

5. $(x-1)^2 + (y-2)^2 = 1$ **6.** $(x+3)^2 + (y+1)^2 = 4$ **7.** $(x-1)^2 + (y+3)^2 = 49$ **8.** $(x+2)^2 + (y-3)^2 = 5$

9. $(x-3)^2 + (y-4)^2 = 9$ **10.** $(x+5)^2 + (y+7)^2 = 2$ **11.** circle, $r = 5$; C=$(3, -1)$ **12.** circle, $r = 2$; C=$(1, 0)$

13. not a circle **14.** circle, $r = 4$; C=$(-2, 3)$ **15.** circle, $r = 4$; C=$(0, 0)$ **16.** circle, $r = 2$; C=$(0, 6)$

17. circle, $r = \sqrt{6}$; C=$(4, 0)$ **18.** circle, $r = 2$; C=$(-1, 3)$ **19.** not a circle **20.** not a circle

2.3 Functions

1. Function, $D = \{-4, -2, 0, 3\}$; $R = \{0, 1, 2, 9\}$ **2.** Not a function **3.** Not a function

4. Function, $D = \{50, 100, 150, 250\}$; $R = \{10, 13, 14\}$ **5.** y is a function of x; $D = (-\infty, \infty)$; $R = [0, \infty)$ **6.** y is a function of x; $D = (-\infty, 2]$; $R = [0, \infty)$

7. y is not a function of x; $D=(-\infty,\infty); R=(-\infty,\infty)$

8. y is a function of x; $D=R=(-\infty,0)\cup(0,\infty)$

9. 27

10. $a^2+2ah+h^2-6a-6h$

11. 0

12. $6t-7$

13. $f(-1)=\dfrac{13}{3}; f(3)=3$

14. $f(-1)=7; f(3)=135$

15. $f(-1)=-1.75; f(3)=5.25$

16. $f(-1)=2; f(3)=0$

17. $f(x)=\dfrac{3}{4}x-\dfrac{5}{4}$

18. $f(x)=2x^2+4$

19. $f(x)=-\dfrac{1}{3}x+2$

20. $f(x)=-6x^2+2$

21. (a) $[0,\infty)$ **(b)** $(-\infty,0]$ **(c)** none

22. (a) $[0,\infty)$ **(b)** none **(c)** none

23. (a) $(-\infty,\infty)$ **(b)** none **(c)** none

24. (a) $[3,\infty)$ **(b)** $(-\infty,3]$ **(c)** none

2.4 Linear Functions

1. $D=R=(-\infty,\infty)$

2. $D=R=(-\infty,\infty)$

3. $D=(-\infty,\infty); R=\{4\}$; constant function

4. $D=R=(-\infty,\infty)$

5. $D=R=(-\infty,\infty)$

6. $D=\{2\}; R=(-\infty,\infty)$

7. $D=\{-3\}; R=(-\infty,\infty)$

8. $D=\{3\}; R=(-\infty,\infty)$

9. $m=3$

10. $m=0$

11. $m=-5$

12. $m=0$

13. undefined

14. $m=-2$

15. $m=1$

16. $m=2$

17. $m=-\dfrac{3}{4}$

18. $m=-3$

19. $m=2$

20. $m=3$

2.5 Equations of Lines; Curve Fitting

1. $2x - 3y = -7$
2. $y = 0$
3. $3x + 5y = 11$
4. $x = 6$
5. $y = 6x + 16$
6. $y = \dfrac{10}{3}x - \dfrac{13}{6}$
7. $x = 2$
8. $y = 6$
9. $m = -\dfrac{1}{2}; y\text{-int} = \dfrac{5}{2}$
10. $m = \dfrac{1}{2}; y\text{-int} = 4$
11. $m = \dfrac{1}{4}; y\text{-int} = -3$
12. $m = \dfrac{3}{2}; y\text{-int} = -2$
13. $2x - y = 0; y = 2x$
14. $3x - 5y = 41; y = \dfrac{3}{5}x - \dfrac{41}{5}$
15. $5x + y = 11; y = -5x + 11$
16. $x = -10$
17. yes
18. yes
19. yes
20. no

2.6 Graphs of Basic Functions

1. $(-\infty, -1] \cup (-1, \infty)$
2. $(-\infty, 0] \cup (0, \infty)$
3. $(-\infty, 3) \cup [3, \infty)$
4. $(-\infty, \infty)$
5. (a) -2, (b) 0, (c) 2, (d) 3
6. (a) 7, (b) 5, (c) 3, (d) 1
7. (a) -3, (b) 6, (c) 5, (d) 0
8. (a) 11, (b) 3, (c) 3, (d) 3
9. (a) -1, (b) -2, (c) 0, (d) -4
10. (a) -14, (b) $-\dfrac{5}{2}$, (c) -2, (d) -2
11. (a) -24, (b) -18, (c) -12, (d) 4
12. (a) 5, (b) 4, (c) 8, (d) 12
13. (a) 17, (b) 11, (c) 5, (d) 10
14. (a) 14, (b) 6, (c) 0, (d) 2
15. (a) -13, (b) -5, (c) 1, (d) 7
16. (a) 16, (b) 8, (c) 3, (d) 4

2.7 Graphing Techniques

1. a vertical stretch by a factor of 3 and a reflection across the x-axis

2. a vertical shrink by a factor of $\frac{1}{3}$

3. a vertical stretch by a factor of 9

4. a vertical stretch by a factor of 2

5. a vertical stretch by a factor of 2 and a reflection across the x-axis

6. a vertical shrink by a factor of $\frac{1}{2}$

7. y-axis

8. y-axis

9. x-axis, y-axis, origin

10. origin

11. none of these

12. y-axis

13. even

14. neither

15. even

16. neither

17. even

18. neither

19. a translation of 1 unit upward

20. a translation of 3 units downward

21. a translation of 1 unit to the left

22. a translation of 3 units to the right

23. a translation of 2 units downward and 1 unit to the right

24. a translation of 2 units to the right

2.8 Function Operations and Composition

1. 26

2. 120

3. 6

4.
$(f+g)(x) = 7x+1;\ D = (-\infty, \infty)$
$(f-g)(x) = 5x+7;\ D = (-\infty, \infty)$
$(f \cdot g)(x) = 6x^2 - 14x - 12;\ D = (-\infty, \infty)$
$\left(\dfrac{f}{g}\right)(x) = \dfrac{6x+4}{x-3};\ D = (-\infty, 3) \cup (3, \infty)$

5.
$(f+g)(x) = 3x^2 - x;\ D = (-\infty, \infty)$
$(f-g)(x) = -x^2 + x - 2;\ D = (-\infty, \infty)$
$(f \cdot g)(x) = 2x^4 - x^3 - x^2 + x - 1;\ D = (-\infty, \infty)$
$\left(\dfrac{f}{g}\right)(x) = \dfrac{x^2 - 1}{2x^2 - x + 1};\ D = (-\infty, \infty)$

36 Answers

6.
$(f+g)(x) = \sqrt{2x+1} + \dfrac{1}{x}; \ D = \left[-\dfrac{1}{2}, 0\right) \cup (0, \infty)$

$(f-g)(x) = \sqrt{2x+1} - \dfrac{1}{x}; \ D = \left[-\dfrac{1}{2}, 0\right) \cup (0, \infty)$

$(f \cdot g)(x) = \dfrac{\sqrt{2x+1}}{x}; \ D = \left[-\dfrac{1}{2}, 0\right) \cup (0, \infty)$

$\left(\dfrac{f}{g}\right)(x) = x\sqrt{2x+1}; \ D = \left[-\dfrac{1}{2}, \infty\right)$

7. (a) 2, (b) 8, (c) 5

8. (a) 4, (b) 4, (c) $\dfrac{1}{3}$

9. (a) 1, (b) 0, (c) undefined

10. (a) $2x + 2h - 1$, (b) $2h$, (c) 2

11. (a) $x^3 + 3x^2h + 3xh^2 + h^3$, (b) $3x^2h + 3xh^2 + h^3$, (c) $3x^2 + 3xh + h^2$

12. (a) $x^2 + 2xh + h^2 - 2$, (b) $2xh + h^2$, (c) $2x + h$

13. 1

14. −15

15. 28

16. $(f \circ g)(x) = 26 - 12x;$
$(g \circ f)(x) = -12x$

17. $(f \circ g)(x) = 6x^2 - 2x + 6;$
$(g \circ f)(x) = 12x^2 + 22x + 12$

18. $(f \circ g)(x) = 5x - 5 + 4\sqrt{x-1};$
$(g \circ f)(x) = \sqrt{5x^2 + 4x - 1}$

3.1 Quadratic Functions and Models

Give the vertex, axis, domain, and range for each of the following parabolas.

1. $f(x) = \left(x - \frac{5}{2}\right)^2 - \frac{1}{4}$

2. $f(x) = (x-3)^2$

3. $f(x) = x^2 - 6x - 1$

4. $f(x) = x^2 + 2x + 4$

5. $f(x) = 2x^2 - 7x + 3$

6. $f(x) = x^2 - 5x + 10$

7. $f(x) = x^2 + x - 6$

8. $f(x) = x^2 - 2x - 3$

9. $f(x) = -2x^2 + 5x + 1$

10. $f(x) = 4x^2 - 12x + 9$

11. $f(x) = 9x^2 - 16$

12. $f(x) = x^2 - 7x + 12$

Modeling

13. A volleyball player hits a ball and it goes straight up into the air with an initial velocity of 16 feet per second from an initial height of 4 feet. Find the function that describes the height of the ball in terms of time t and determine the time at which the ball reaches its maximum height.

14. Find two numbers whose sum is 50 and whose product is a maximum.

15. A farmer has 36 feet of fence to build a pigpen. He is going to use one of the sides of his barn as a side to the rectangular enclosure. Determine a function A that represents the total area of the enclosed region. What is the maximum area that can be enclosed?

16. A square piece of poster board is going to be made into an open box by cutting 3-inch squares from each corner and folding up the sides. Determine a function V that represents the volume of the box.

17. A gardener wants to dig up a piece of the yard and make a rectangular vegetable garden with a border around the four sides and a piece of border splitting the region in half. She has 48 feet of border. Write a function A that represents the area of the garden. What is the maximum area that can be enclosed?

18. A toy rocket is launched into the sky with an initial velocity of 193 feet per second from a launching pad 6 feet in the air. Find the function that describes the height of the rocket in terms of time t and determine the time at which the ball reaches its maximum height.

19. A math book is thrown straight up from the ground with an initial velocity of 96 feet per second. Find the function that describes the height of the book in terms of time t and determine the time at which the book reaches its maximum height.

20. Suppose that x represents one of two positive numbers whose sum is 30. Determine a function f that represents the product of these two numbers. For what number(s) is the product a maximum?

3.2 Synthetic Division

Use synthetic division to perform each division.

1. $\dfrac{5x^3 - 3x + 4}{x - 2}$

2. $\dfrac{2x^3 - 3x^2 + 1}{x - 3}$

3. $\dfrac{x^3 + 4x^2 - 7}{x + 3}$

4. $\dfrac{3x^4 - 4x^2 + 4}{x + 2}$

5. $\dfrac{2x^3 + 3x^2 + 4x + 7}{x - \frac{1}{2}}$

6. $\dfrac{x^3 + x}{x + 1}$

Express *f(x)* in the form $f(x) = (x-k) \cdot q(x) + r$ for the given value of *k*.

7. $f(x) = x^2 - x - 6;\ k = 3$

8. $f(x) = 12x^2 - 20x + 3;\ k = \dfrac{3}{2}$

9. $f(x) = x^2 + 11x + 16;\ k = -8$

10. $f(x) = 5x^4 + 5x^3 + 2x^2 - x;\ k = -1$

11. $f(x) = 2x^3 - 5x^2 - 6x + 17;\ k = 3$

12. $f(x) = 3x^3 - 4x^2 - 6x + 7;\ k = 2$

For each polynomial function use the remainder theorem and synthetic division to find *f(k)*.

13. $k = -1;\ f(x) = x^3 - x^2 + 4x + 2$

14. $k = -\dfrac{1}{2};\ f(x) = -8x^5 - 7x^4 + 9x$

15. $k = 1;\ f(x) = 7x^4 - 4x^3 + 1$

16. $k = 4;\ f(x) = x^3 - 2x^2 - 7x + 2$

17. $k = 3;\ f(x) = 6x^3 + 8x^2 - 17x + 6$

18. $k = -5;\ f(x) = 6x^2 - 9x - 6$

Use synthetic division to decide whether the given number *k* is a zero of the given polynomial.

19. $f(x) = x^3 + 3x^2 - x - 3;\ k = -1$

20. $f(x) = x^4 - 8x^3 + 22x^2 - 24x + 9;\ k = 2$

21. $f(x) = x^4 + 3x^3 - x^2 - 3x;\ k = -3$

22. $f(x) = x^3 + 6x^2 - x - 30;\ k = -5$

23. $f(x) = x^5 - 4x^3;\ k = 3$

24. $f(x) = x^4 + x^3 - 8x^2 - 2x + 12;\ k = 2$

3.3 Zeros of Polynomial Functions

Use the factor theorem and synthetic division to decide whether the second polynomial is a factor of the first.

1. $5x^4 + 7x^3 - 6x^2 + 4x + 8$; $x+2$
2. $x^3 - 5x^2 - 2x + 24$; $x-3$
3. $8x^3 - 36x^2 + 54x - 27$; $x-3$
4. $x^3 - 6x^2 + 11x - 6$; $x+1$

Factor $f(x)$ into linear factors given that k is a zero of $f(x)$.

5. $f(x) = x^3 - x^2 - 24x - 36$; $k = 6$
6. $f(x) = x^3 - 6x^2 - x + 30$; $k = 3$
7. $f(x) = x^3 - 9x^2 + 2x + 48$; $k = -2$
8. $f(x) = x^3 - 6x^2 + 3x + 10$; $k = -1$

For each polynomial function, one zero is given. Find all others.

9. $f(x) = 2x^3 + 3x^2 - 2x$; -2
10. $f(x) = x^3 + 4x^2 - 10x + 12$; $1+i$
11. $f(x) = x^3 - 5x^2 - 16x + 80$; 5
12. $f(x) = 2x^3 + 3x^2 + 2x + 3$; i

For each polynomial function, find all rational zeros and factor $f(x)$.

13. $f(x) = x^3 + 7x^2 - 28x + 20$
14. $f(x) = x^3 - 12x^2 - 55x + 150$
15. $f(x) = 5x^3 + 36x^2 - 33x - 8$
16. $f(x) = 4x^3 + 9x^2 - 81x + 54$

For each polynomial function, find all zeros and their multiplicities.

17. $f(x) = x^3 - 3x^2 - 9x - 5$
18. $f(x) = x^3 - 6x^2 + 32$
19. $f(x) = x(x - 3 + \sqrt{17})(5x + 1)$
20. $f(x) = (5x - 8)^4 (3x + 2)^2$

Find a polynomial function of degree 3 with real coefficients that satisfies the given conditions.

21. zeros of 6, -5, 8; $f(1) = 420$
22. zeros of 1, 4, 10; $f(2) = 48$
23. zeros of -3, -4, 1; $f(-1) = 12$
24. zeros of 10, -2, 1; $f(-2) = 0$

Find a polynomial function of least degree having only real coefficients with zeros as given.

25. $1-i$ and $2+i$
26. $4+i$ and $5-i$

3.4 Polynomial Functions: Graphs, Applications, and Models

Describe the end behavior of the graph of each polynomial function.

1. $f(x) = 3x^3 - 2x^2 + x - 1$

2. $f(x) = 4x^6 - 5x^3 - 2x$

3. $f(x) = x^4 + 2x^3 - 6x^2 + 2$

4. $f(x) = 2x^5 + x^4 - 6x^3$

Use the intermediate value theorem for polynomials to show that each polynomial function has a real zero between the numbers given.

5. $f(x) = 3x^3 + 5x^2 - 5x + 1$; -3 and 2

6. $f(x) = 8x^3 - 6x^2 - 5x + 3$; -1 and 0

7. $f(x) = 4x^3 + 9x^2 - 10x + 3$; -4 and -3

8. $f(x) = 6x^4 + 4x^3 + 4x - 6$; 0 and 1

Show that the real zeros of each polynomial function satisfy the given conditions.

9. $f(x) = 2x^3 + 10x^2 - 2x$; no real zeros greater than 1

10. $f(x) = 3x^3 + 18x^2 - 24x$; no real zeros greater than 2

11. $f(x) = 8x^3 + 2x - 30$; no real zeros less than 1

12. $f(x) = 27x^3 + 3x + 68$; no real zeros greater than -1

For the given polynomial function, approximate each zero as a decimal to the nearest tenth.

13. $f(x) = 2x^3 + 3x^2 + 4x - 6$

14. $f(x) = 4x^6 + 8x^5 + 2x^4 - 3x^3 - 5x^2 - 6x - 9$

15. $f(x) = 5x^3 + 9x^2 - 10x - 17$

16. $f(x) = 5x^4 - 6x^3 + 8x - 9$

Use a graphing calculator to find the coordinates of the turning points of the graph of each polynomial function in the given domain interval. Give answers to the nearest hundredth.

17. $f(x) = x^3 + 4x^2 - 23x - 35$; $[-1, 4]$

18. $f(x) = 2x^4 + 2x^3 - 3x^2 - 6$; $[0.1, 1]$

19. $f(x) = 5x^3 - 3x^2 + 7$; $[0.2, 1]$

20. $f(x) = x^4 - 3x^3 + 2x^2 - 1$; $[1, 2]$

Explain how the graph of each function can be obtained from the graph of $y=\dfrac{1}{x}$ or $y=\dfrac{1}{x^2}$.

1. $y = -\dfrac{4}{x^2}$

2. $y = \dfrac{1}{x-2}$

3. $y = \dfrac{1}{x} - 7$

4. $y = \dfrac{1}{x^2} + 2$

5. $y = \dfrac{1}{x-6}$

6. $y = \dfrac{1}{(x+2)^2}$

7. $y = \dfrac{1}{x-3} + 4$

8. $y = \dfrac{3}{(x+4)^2} - 2$

Give the equations of any vertical, horizontal, or oblique asymptotes for the graph of each rational function.

9. $y = \dfrac{1}{x-4}$

10. $y = \dfrac{5}{x^2 - x - 6}$

11. $y = \dfrac{2x}{2x+4}$

12. $y = \dfrac{x+3}{x-1}$

13. $y = \dfrac{x^2 + 9}{x^2 - 4}$

14. $y = \dfrac{x^2 - 9}{x - 1}$

15. $y = \dfrac{3x^2}{2x^2 + 4}$

16. $y = \dfrac{x^2 + 11x + 16}{x + 8}$

Find a possible equation for the function with a graph having the given features.

17.
  x-intercept: 1   y-intercept: $-\dfrac{1}{2}$
  V.A.: $x = -2$ H.A.: $y = 1$

18.
  x-intercept: 0   y-intercept: 0
  V.A.: $x = 4$ H.A.: $y = 1$

3.6 Variation

Solve each variation problem.

1. If y varies directly as x, and $y = 30$ when $x = 8$, find y when $x = 4$.

2. Two yards of fabric cost $10.52. How much will 6.5 yards of fabric cost?

3. The distance a car can travel varies directly with the amount of gas in the tank. On the highway, a small SUV travels 212 miles on 13 gallons of gas. How many gallons of gas will be used on a 1036-mile trip?

4. The length, L, of a pendulum varies directly with the square of its period, p, the time required for the pendulum to make a complete swing back and forth. The pendulum on a clock is 6 inches long and has a period of 2 seconds. If a clock has a pendulum that is 8 inches long, how long is its period?

5. The distance a projectile will travel varies directly as the square of the initial velocity. A projectile with an initial velocity of 300 feet per second travels for 15,000 feet. What must the initial velocity be in order for the projectile to travel 60,000 feet?

6. If y varies inversely as x, and $y = 3$ when $x = 4$, find y when $x = 8$.

7. If y varies inversely as $\sqrt{x}$, and $y = \frac{1}{4}$ when $x = 4$, find y when $x = 169$.

8. The time in days, d, required to relandscape a park is inversely proportional to the number of people, n, landscaping. If 15 people can do the job in 4 days, how many days would it take 6 people to do the job?

9. The amount of current, C, that flows through a circuit varies inversely with the resistance, R, on the circuit. A rod with a resistance of 6 ohms draws 5 amps of current. Find the current when the resistance is 500 ohms.

10. The intensity of illumination, I, from a lamp varies inversely with the square of the distance, d, the object is from the lamp. If the intensity at 5 feet is 48 foot-candles, find the intensity at 6 feet.

11. If y is jointly proportional to x and the square root of z, and $y = -18.954$ when $x = -1.35$ and $z = 15.21$, find y when $x = 4$ and $z = 144$.

12. If y varies directly as x and inversely as z^2, and $y = 100$ when $x = 1$ and $z = 3$, find y when $x = 5$ and $z = 2$.

3.1 Quadratic Functions and Models

1. Vertex: $(2.5, -.25)$ Axis: $x = 2.5$ Domain: $(-\infty, \infty)$ Range: $[-.25, \infty)$

2. Vertex: $(3, 0)$ Axis: $x = 3$ Domain: $(-\infty, \infty)$ Range: $[0, \infty)$

3. Vertex: $(3, -10)$ Axis: $x = 3$ Domain: $(-\infty, \infty)$ Range: $[-10, \infty)$

4. Vertex: $(-1, 3)$ Axis: $x = -1$ Domain: $(-\infty, \infty)$ Range: $[3, \infty)$

5. Vertex: $(1.75, -3.125)$ Axis: $x = 1.75$ Domain: $(-\infty, \infty)$ Range: $[-3.125, \infty)$

6. Vertex: $(2.5, 3.75)$ Axis: $x = 2.5$ Domain: $(-\infty, \infty)$ Range: $[3.75, \infty)$

7. Vertex: $(-.5, -6.25)$ Axis: $x = -.5$ Domain: $(-\infty, \infty)$ Range: $[-6.25, \infty)$

8. Vertex: $(1, -4)$ Axis: $x = 1$ Domain: $(-\infty, \infty)$ Range: $[-4, \infty)$

9. Vertex: $(1.25, 4.125)$ Axis: $x = 1.25$ Domain: $(-\infty, \infty)$ Range: $(-\infty, 4.125]$

10. Vertex: $(1.5, 0)$ Axis: $x = 1.5$ Domain: $(-\infty, \infty)$ Range: $[0, \infty)$

11. Vertex: $(0, -16)$ Axis: $x = 0$ Domain: $(-\infty, \infty)$ Range: $[-16, \infty)$

12. Vertex: $(3.5, -.25)$ Axis: $x = 3.5$ Domain: $(-\infty, \infty)$ Range: $[-.25, \infty)$

13. $h(t) = -16t^2 + 16t + 4$; maximum height after 0.5 second

14. 25 and 25

15. $A(x) = x(36 - 2x)$; maximum area of 162 square feet

16. $V(x) = 3(x - 6)^2$

17. $A(x) = x(24 - \frac{3}{2}x)$; maximum area of 96 square feet

18. $h(t) = -16t^2 + 193t + 6$; maximum height after 6.03 seconds

19. $h(t) = -16t^2 + 96t$; maximum height after 3 seconds

20. $f(x) = x(30 - x)$; $x = 15$

Answers

3.2 Synthetic Division

1. $5x^2 + 10x + 17 + \dfrac{38}{x-2}$
2. $2x^2 + 3x + 9 + \dfrac{28}{x-3}$
3. $x^2 + x - 3 + \dfrac{2}{x+3}$
4. $3x^3 - 6x^2 + 8x - 16 + \dfrac{36}{x+2}$
5. $2x^2 + 4x + 6 + \dfrac{10}{x - \frac{1}{2}}$
6. $x^2 - x + 2 - \dfrac{2}{x+1}$
7. $f(x) = (x-3)(x+2)$
8. $f(x) = (x - \dfrac{3}{2})(12x - 2)$
9. $f(x) = (x+8)(x+3) - 8$
10. $f(x) = (x+1)(5x^3 + 2x - 3) + 3$
11. $f(x) = (x-3)(2x^2 + x - 3) + 8$
12. $f(x) = (x-2)(3x^2 + 2x - 2) + 3$
13. -4
14. $-\dfrac{75}{16}$
15. 4
16. 6
17. 189
18. 189
19. yes
20. no
21. yes
22. yes
23. no
24. yes

3.3 Zeros of Polynomial Functions

1. yes
2. yes
3. no
4. no
5. $(x-6)(x+2)(x+3)$
6. $(x-5)(x-3)(x+2)$
7. $(x-8)(x-3)(x+2)$
8. $(x-5)(x-2)(x+1)$
9. $x = \dfrac{1}{2}, 0, -2$
10. $x = 1+i, 1-i, -6$
11. $x = 5, 4, -4$
12. $x = i, -i, -3/2$
13. $(x-2)(x-1)(x+10)$
14. $(x-15)(x-2)(x+5)$
15. $(x-1)(x+8)(5x+1)$
16. $(x-3)(x+6)(4x-3)$
17. $x = 5$ multiplicity 1; $x = -1$ multiplicity 2
18. $x = -2$ multiplicity 1; $x = 4$ multiplicity 2
19. $x = 0, 3-\sqrt{17}, -1/5$

20. $x = \dfrac{8}{5}$ multiplicity 4; $x = -\dfrac{2}{3}$ multiplicity 2

21. $f(x) = 2x^3 - 18x^2 - 44x + 480$

22. $f(x) = 3x^3 - 45x^2 + 162x - 120$

23. $f(x) = -x^3 - 6x^2 - 5x + 12$

24. $f(x) = x^3 - 9x^2 - 12x + 20$

25. $f(x) = x^4 - 6x^3 + 15x^2 - 18x + 10$

26. $f(x) = x^4 - 18x^3 + 123x^2 - 378x + 442$

3.4 Polynomial Functions: Graphs, Applications, and Models

1. as $x \to -\infty, f(x) \to -\infty$, and as $x \to \infty, f(x) \to \infty$

2. as $x \to -\infty, f(x) \to \infty$, and as $x \to \infty, f(x) \to \infty$

3. as $x \to -\infty, f(x) \to \infty$, and as $x \to \infty, f(x) \to \infty$

4. as $x \to -\infty, f(x) \to -\infty$, and as $x \to \infty, f(x) \to \infty$

5. $f(-3) = -20 < 0$; $f(2) = 35 > 0$

6. $f(-1) = -6 < 0$; $f(0) = 3 > 0$

7. $f(-4) = -69 < 0$; $f(-3) = 6 > 0$

8. $f(0) = -6 < 0$; $f(1) = 8 > 0$

9. zeros = .19, 0, −5.19

10. zeros = 1.12, 0, −7.12

11. zero = 1.5

12. zero = −1.33

13. zero = .8

14. zeros = 1.1, −1.7

15. zeros = 1.4, −1.3, −1.9

16. zeros = 1.2, −1.2

17. (1.74, −57.64)

18. (.57, −6.39)

19. (.4, 6.84)

20. (1.64, −1.62)

46 Answers

3.5 Rational Functions: Graphs, Applications, and Models

1. a vertical stretch by a factor of 4 and a reflection across the *x*-axis

2. a shift of 2 units to the right

3. a shift of 7 units down

4. a shift of 2 units up

5. a shift of 6 units to the right

6. a shift of 2 units to the left

7. a shift of 3 units to the right and a shift of 4 units up

8. a vertical stretch by a factor of 3, a shift of 4 units to the left and a shift of 2 units down

9. V.A.: $x=4$; H.A.: $y=0$

10. V.A.: $x=3$ and $x=-2$; H.A.: $y=0$

11. V.A.: $x=-2$; H.A.: $y=1$

12. V.A.: $x=1$; H.A.: $y=1$

13. V.A.: $x=2$ and $x=-2$; H.A.: $y=1$

14. V.A.: $x=1$; O.A.: $y=x+1$

15. V.A.: none; H.A.: $y=1.5$

16. V.A.: $x=-8$; O.A.: $y=x+3$

17. $y=\dfrac{x-1}{x+2}$

18. $y=\dfrac{x}{x-4}$

3.6 Variation

1. $y=15$

2. $34.19

3. 63.5 gallons

4. 2.3 seconds

5. 600 feet per second

6. $y=\dfrac{3}{2}$

7. $y=\dfrac{1}{26}$

8. 10 days

9. 0.06 amps

10. 33.3 foot-candles

11. 172.8

12. 1125

Decide whether each function as defined is one-to-one.

1. $f(x) = 3x + 6$

2. $f(x) = \dfrac{1}{x^2 - 1}$

3. $f(x) = x^2 - 5$

4. $f(x) = \sqrt[3]{x - 1}$

5. $f(x) = \sqrt{x - 5}$

Determine whether each pair of functions defined as follows are inverses of each other.

6. $f(x) = 4x + 10;\ g(x) = \dfrac{1}{4}x - \dfrac{5}{2}$

7. $f(x) = x^2 - 2;\ g(x) = \sqrt{x + 2}$

8. $f(x) = x^3 - 4;\ g(x) = \sqrt[3]{x + 4}$

9. $f(x) = \dfrac{1}{x^2 - 3};\ g(x) = -\sqrt{\dfrac{1}{x} + 3}$

10. $f(x) = \dfrac{3}{x - 1};\ g(x) = \dfrac{3}{x} + 1$

For each function defined as follows that is one-to-one, write an equation for the inverse function.

11. $f(x) = 3x + 9$

12. $f(x) = 3x^2 - 6x$

13. $f(x) = x^3 - 1$

14. $f(x) = \dfrac{1}{x + 6}$

15. $f(x) = \dfrac{1}{x^2} - 2$

16. $f(x) = \sqrt{x - 2}$

17. $f(x) = (6x - 1)^2$

18. $f(x) = \sqrt[3]{x} + 5$

19. $f(x) = (x + 1)^3$

20. $f(x) = \sqrt{5 - x}$

4.2 Exponential Functions

If $f(x) = 2^x$ and $g(x) = \left(\dfrac{1}{3}\right)^x$, find each of the following.

1. $f(1)$
2. $f(-2)$
3. $f(6.47)$
4. $g(0)$
5. $g(-1)$
6. $g(0.61)$

Solve each equation.

7. $2^{x+2} = 8^{4/3}$
8. $9 \cdot 3^{x-3} = 27^{-2x}$
9. $2^{x^2-x-4} = 4$
10. $25^{2-3x} = 125^{x-5}$
11. $\left(\dfrac{1}{10}\right)^x = 1000^{2x+4}$
12. $e^x = e^{6x-9}$
13. $\sqrt{e} = e^{2x}$
14. $x^{2/3} = 11$

Solve each problem involving compound interest.

16. Find the future value and interest earned if $13,562 is invested at 2% compounded quarterly for 10 years.

17. Find the present value of $8000 if interest is 4% compounded semi-annually for 2 years.

18. Find the required annual interest rate for $5000 to grow to $5261, if interest is compounded monthly for 12 months.

19. How long will it take for $3000 to grow to $4000 at an interest rate of 6% compounded continuously?

Use a graphing calculator to find the solution set of each equation.

20. $3^{4x-1} = x + 2$

21. $e^{2x+1} = x$

22. $x = 3^x - 3$

4.3 Logarithmic Functions

For each statement, write an equivalent statement in logarithmic or exponential form.

1. $\log_{10} 0.01 = -2$

2. $9^{1/2} = 3$

Solve each logarithmic equation.

3. $x = \log_2 \frac{1}{4}$

4. $\log_{10} 1 = x$

5. $x = 3^{\log_3 27}$

6. $\log_x 64 = 6$

7. $\log_5 x = 2$

8. $x = \log_2 \sqrt[3]{32}$

Write each expression as a sum, difference, or product of logarithms. Simplify the result if possible. Assume all variables represent positive real numbers.

9. $\log_2 \frac{x}{yz}$

10. $\log_b \sqrt{\frac{x^2}{y^3}}$

11. $\log_m \sqrt[3]{\frac{x^2 y^5}{z^7}}$

12. $\log_7 (x^2 - 2y^2)$

Write each expression as a single logarithm with coefficient 1. Assume that all variables represent positive real numbers.

13. $2\log_2 x + 2\log_2 y - \frac{1}{2}\log_2 z$

14. $\frac{1}{3}\log_m x - \frac{1}{4}\log_m y$

15. $\log_2 (x^2 - 9) - \log_2 (x - 3)$

Given $\log_{10} 3 = 0.4771$ and $\log_{10} 7 = 0.8451$, find each logarithm without using a calculator.

16. $\log_{10} 21$

17. $\log_{10} \left(\frac{7}{3}\right)$

18. $\log_{10} 63$

19. $\log_{10} \left(\frac{27}{49}\right)$

4.4 Evaluating Logarithms and the Change-of-Base Theorem

Use a calculator with logarithm keys to find an approximation for each expression. Give answers to four decimal places.

1. $\log(14)$

2. $\log(0.196)$

3. $\ln(30.6)$

4. $\ln(3 \cdot e^2)$

For each substance, find the pH from the given hydronium ion concentration.

5. ammonia, 3.16×10^{-12}

6. battery acid, 3.16×10^{-1}

Find the $[H_3O^+]$ for each substance with the given pH.

7. cola, 2.5

8. bleach, 12.5

Use the change-of-base theorem to find an approximation for each logarithm. Give answers to 4 decimal places.

9. $\log_3 24$

10. $\log_2 3$

11. $\log_8 2.7$

12. $\log_{\sqrt{2}} 9$

Given $f(x) = e^x$; $g(x) = 2^x$; $h(x) = \ln x$; $j(x) = \log_5 x$, **evaluate the following.**

13. $f(\ln 6)$

14. $g(\log_2 \sqrt{3})$

15. $h(e^{3\ln 5})$

16. $j(5^{2\log_5 5})$

Use the properties of logarithms to describe how the graph of the given function compares to the graph of $f(x) = \ln x$.

17. $f(x) = \ln(e^{1/3} x)$

18. $f(x) = \ln(x^{-1})$

19. $f(x) = \ln(x) - \ln(e^4)$

20. $f(x) = \ln(x^3)$

4.5 Exponential and Logarithmic Equations

Solve each equation. Round answers to 4 decimal places.

1. $10^{x+7} = 13$

2. $2^{2x-1} = 9$

3. $25^x = 10^{-3x}$

4. $e^{2x} = 4e^x$

5. $4^{2x+1} = 6^x$

6. $(x-2)e^{-x} = 0$

7. $\log_2(x+1) + \log_2(x-3) = 1$

8. $\log_3(x+2) - \log_3(x) = 3$

9. $\log x + \log(x-4) = 6$

10. $3\log x - 5 = 4$

11. $2\ln(4x) = 16$

12. $\log(x^2 - 9) - \log(x+3) = 1$

Solve each equation for the indicated variable. Use logarithms to the appropriate bases.

13. $N = 225e^{0.02t}$ for t

14. $y = -301\ln\left(\dfrac{x}{207}\right)$ for x

15. $R = 10 \cdot 2^{-t/5700}$ for t

16. $y = Ce^{kt}$ for t

Use a graphing calculator to solve each equation. Give irrational solutions correct to the nearest hundredth.

17. $x - 4 = \ln x$

18. $\ln(3 - e^x) = 5$

19. $\log x = 0.5x - 2$

20. $4 = x + \log_4 x$

4.6 Applications and Models of Exponential Growth and Decay

Solve each problem.

1. The population of a town is $P = 95,700e^{0.013t}$ where $t = 0$ represents 1999. When will the population reach 200,000?

2. The number of flies in a lab is $N = 75e^{0.03t}$ where t is the number of days. When will the number of flies double?

3. The number of ants in a colony is $N = 100e^{0.012t}$ where t is the number of days. When will there be 600 ants?

4. A plasma TV depreciates according to $P = 3600e^{-0.19t}$ where t is the age of the plasma TV in years. How much will the TV be worth in 5 years?

5. The magnitude, M, of an earthquake is measured with the Richter scale, $M = \log_{10} I$ where I is the intensity of the earthquake. Find the intensity of the 1971 earthquake in Los Angeles that had a magnitude of 6.7.

6. The level of sound is measured in decibels and is modeled by the formula $D = 10 \log_{10}\left(\dfrac{I}{10^{-12}}\right)$ where I is the intensity of the sound in watts per square meter. Find the decibel level for a noise with an intensity of $10^{-2.5}$ watt per square meter.

7. Refer back to exercise 6. What is the intensity of a sound with a decibel level of 89?

8. Find the doubling time of an investment earning 6% interest if interest is compounded continuously.

9. How long will it take $30,000 to grow to $75,000 if it is invested at 5.5% interest compounded continuously?

10. Find the amount of money you would need to invest now at 8% compounded continuously to have $100,000 in 10 years.

11. Find the tripling time of an investment earning 9% interest compounded continuously.

12. What interest rate would give you $90,000 if you invest $50,000 for 20 years in an account that compounds interest continuously?

4.1 Inverse Functions

1. yes

2. no

3. no

4. yes

5. yes

6. yes

7. no

8. yes

9. no

10. yes

11. $f^{-1}(x) = \frac{1}{3}x - 3$

12. not one-to-one

13. $f^{-1}(x) = \sqrt[3]{x+1}$

14. $f^{-1}(x) = \frac{1}{x} - 6$

15. not one-to-one

16. $f^{-1}(x) = x^2 + 2; x \geq 0$

17. not one-to-one

18. $f^{-1}(x) = (x-5)^3$

19. $f^{-1}(x) = \sqrt[3]{x} - 1$

20. $f^{-1}(x) = 5 - x^2; x \geq 0$

4.2 Exponential Functions

1. 2

2. $\frac{1}{4}$

3. 88.647

4. 1

5. 3

6. 0.51163

7. 2

8. $\frac{1}{7}$

9. −2, 3

10. $\frac{19}{9}$

11. $-\frac{12}{7}$

12. $\frac{9}{5}$

13. $\frac{1}{4}$

14. $-11^{3/2}, 11^{3/2}$

15. $16,556.40, $2994.41

16. $7390.76

17. 5.1%

18. 4.8 years

19. 0.45, −2

20. no solution

21. −2.96, 1.34

22. 0.56

54 Answers

4.3 Logarithmic Functions

1. $10^{-2} = 0.01$

2. $\log_9 3 = \dfrac{1}{2}$

3. -2

4. 0

5. 27

6. 2

7. 25

8. $\dfrac{5}{3}$

9. $\log_2 x - \log_2 y - \log_2 z$

10. $\log_b x - \dfrac{3}{2}\log_b y$

11. $\dfrac{2}{3}\log_m x + \dfrac{5}{3}\log_m y - \dfrac{7}{3}\log_m z$

12. $\log_7(x^2 - 2y^2)$

13. $\log_2\left(\dfrac{x^2 y^2}{\sqrt{z}}\right)$

14. $\log_m\left(\dfrac{\sqrt[3]{x}}{\sqrt[4]{y}}\right)$

15. $\log_2(x+3)$

16. 1.3222

17. 0.3680

18. 1.7993

19. -0.2589

4.4 Evaluating Logarithms and the Change-of-Base Theorem

1. 1.1461

2. -0.7077

3. 3.4210

4. 3.0986

5. 11.5

6. 0.5

7. 3.16×10^{-3}

8. 3.16×10^{-13}

9. 2.8928

10. 1.5850

11. 0.4777

12. 6.3399

13. 6

14. $\sqrt{3}$

15. $3\ln 5$

16. 2

17. shifted up $\dfrac{1}{3}$ unit

18. reflection about the x-axis

4.5 Exponential and Logarithmic Equations

1. −5.8861
2. 2.0850
3. 0
4. 1.3863
5. −1.4134
6. 2
7. 3.4495
8. 0.0769
9. 1002.0020
10. 1000
11. 745.2395
12. 13
13. $t = 50\ln\left(\dfrac{N}{225}\right)$
14. $x = 207e^{-y/301}$
15. $t = -5700\log_2\left(\dfrac{R}{10}\right)$
16. $t = \dfrac{1}{k}\ln\left(\dfrac{y}{C}\right)$
17. 5.75, 0.02
18. no solution
19. 5.48, 0.01
20. 3.17

4.6 Applications and Models of Exponential Growth and Decay

1. 2055
2. 23 days
3. 149 days
4. $1392.27
5. 5.01×10^6
6. 95 decibels
7. 7.9×10^{-4} w/m^2
8. 11.55 years
9. 16.7 years
10. $44,932.90
11. 12.2 years
12. 2.9%

5.1 Angles

Find the complement and the supplement of each angle.

1. 80°
2. 10°
3. 40°
4. 9°

Find the measure of each angle.

5. supplementary angles with measures $10x$ and $5x$ degrees

6. supplementary angles with measures $6x + 4$ and $2x$ degrees

7. complementary angles with measures $7x - 9$ and $7x + 15$ degrees

8. complementary angles with measures $10x$ and $20x$ degrees

Perform each calculation.

9. $135°16'45'' + 234°50''$

10. $41°47'42'' - 32°15'9''$

11. $26°47' - 31°10'$

12. $42°46'15'' + 25°7'37''$

Convert each angle measure to decimal degrees. Round to the nearest thousandth of a degree.

13. $26°30'$

14. $66°54'3''$

15. $10°42'5''$

16. $121°51'22''$

Convert each angle measure to degrees, minutes, and seconds.

17. 73.2°

18. 48.27°

19. 83.41°

20. -36.78°

Find the angle of smallest positive measure coterminal with each angle.

21. -60°

22. 470°

23. 1500°

24. -123°

Give an expression that generates all angles coterminal with each angle. Let n represent any integer.

25. 60°

26. 120°

5.2 Trigonometric Functions

Find the values of the six trigonometric functions for angles in standard position having the following point on their terminal side.

1. $(4,-3)$
2. $(-6,0)$
3. $(-\sqrt{3},1)$

In the following exercises, an equation of the terminal side of an angle θ in standard position is given with a restriction on x. Find the values of the six trigonometric functions of θ.

4. $-12x+5y=0;\ x\geq 0$
5. $-x+y=0;\ x\leq 0$
6. $40x+9y=0;\ x\geq 0$

Use the trigonometric function values of quadrantal angles to evaluate each expression.

7. $\sin^2 90° + \cos 180° \tan 0°$
8. $\cos 90° \sin 90° - \tan^2 0°$
9. $\csc^2 45° \sin^2 315°$

Use the appropriate reciprocal identity to find each function value.

10. $\sin\theta$ if $\csc\theta = -2$
11. $\cos\theta$ if $\sec\theta = \sqrt{11}$
12. $\tan\theta$ if $\cot\theta = \dfrac{\sqrt{6}}{2}$

Identify the quadrant for the angle satisfying the given conditions.

13. $\sin\theta < 0,\ \tan\theta < 0$
14. $\csc\theta > 0,\ \cot\theta < 0$
15. $\sec\theta < 0,\ \sin\theta > 0$

Decide whether each statement is possible or impossible for an angle θ.

16. $\csc\theta = 0.5$
17. $\tan\theta = -6$
18. $\cos\theta = 1$

Use identities to find each function value.

19. $\tan\theta$, if $\cos\theta = -\dfrac{4}{5}$, with θ in quadrant III

20. $\cos\theta$, if $\cot\theta = -1$, with θ in quadrant II

21. $\sin\theta$, if $\tan\theta = 2$, with θ in quadrant I

Find all the trigonometric function values for each angle.

22. $\cos\theta = -\dfrac{8}{17}$, with θ in quadrant II

23. $\sec\theta = \dfrac{13}{5}$, with θ in quadrant IV

5.3 Evaluating Trigonometric Functions

For each expression, give the exact value.

1. $\cot 60°$
2. $\sin 60°$
3. $\sec 60°$

Find the reference angle for each of the following angles.

4. $127°$
5. $398°$
6. $-116°$

Find exact values of the six trigonometric functions for each angle.

7. $135°$
8. $390°$
9. $-690°$

Find the exact value of each expression.

10. $\sin 480°$
11. $\cos 510°$
12. $\tan(-1950°)$

Determine whether each statement is true or false.

13. $\sin 60° = 2 \sin 30° \cos 30°$
14. $\cos^2 135° - \sin^2 135° = 1$
15. $\cos 90° = \cos 60° + \cos 30°$

Use a calculator to find a decimal approximation for each value.

16. $\sec 238°$
17. $\csc 253.8°$
18. $\cot 101.7°$

Find a value of θ in $[0°, 90°]$ that satisfies each statement.

19. $\sin \theta = 0.766$
20. $\cos \theta = 0.309$
21. $\tan \theta = 0.2644$

Find all values of θ in $[0°, 360°)$ that has the given function values.

22. $\sin \theta = -\dfrac{\sqrt{3}}{2}$
23. $\cos \theta = \dfrac{1}{2}$
24. $\tan \theta = 1$

Solve each right triangle. In each case, $C = 90°$.

1. $A = 30°$, $b = 100$
2. $B = 35°$, $b = 5$
3. $A = 32.6°$, $a = 41.6$
4. $A = 50°$, $c = 12$
5. $B = 40°$, $c = 9.1$
6. $B = 32.54°$, $a = 6.7$
7. $B = 52°$, $a = 31$
8. $A = 75°$, $c = 20$
9. $A = 68°$, $b = 80$
10. $b = 21.12$, $c = 52.25$
11. $B = 30°$, $a = 40$
12. $a = 15$, $b = 20$
13. $B = 38°$, $a = 15$
14. $c = 6.587$, $a = 5.476$
15. $b = 120$, $c = 140$
16. $a = 28.4$, $c = 51.3$
17. $A = 60°$, $c = 22$
18. $B = 25.4°$, $c = 3.24$
19. $A = 78°$, $b = 200$
20. $a = 30$, $b = 82.42$

60   Answers

5.1 Angles

1. 10°, 100°   **2.** 80°, 170°   **3.** 50°, 140°

4. 81°, 171°   **5.** 120°, 60°   **6.** 136°, 44°

7. 33°, 57°   **8.** 60°, 30°   **9.** 369°17′35″

10. 9°32′33″   **11.** −4°23′   **12.** 67°53′52″

13. 26.500°   **14.** 66.901°   **15.** 10.701°

16. 121.911°   **17.** 73°12′   **18.** 48°16′12″

19. 83°24′36″   **20.** −36°46′48″   **21.** 300°

22. 110°   **23.** 60°   **24.** 237°

25. $60° + 360n°$ **26.** $120° + 360n°$

5.2 Trigonometric Functions

In exercises 1-6 and 22-23, we give, in order sine, cosine, tangent, cotangent, secant, and cosecant.

1. $-\dfrac{3}{5}; \dfrac{4}{5}; -\dfrac{3}{4}; -\dfrac{4}{3}; \dfrac{5}{4}; -\dfrac{5}{3}$ **2.** 0; −1; 0; undefined; −1; undefined

3. $\dfrac{1}{2}; -\dfrac{\sqrt{3}}{2}; -\dfrac{\sqrt{3}}{3}; -\sqrt{3}; -\dfrac{2\sqrt{3}}{3}; 2$ **4.** $\dfrac{12}{13}; \dfrac{5}{13}; \dfrac{12}{5}; \dfrac{5}{12}; \dfrac{13}{5}; \dfrac{13}{12}$

5. $-\dfrac{\sqrt{2}}{2}; -\dfrac{\sqrt{2}}{2}; 1; 1; -\sqrt{2}; -\sqrt{2}$ **6.** $-\dfrac{40}{41}; \dfrac{9}{41}; -\dfrac{40}{9}; -\dfrac{9}{40}; \dfrac{41}{9}; -\dfrac{41}{40}$

7. 1   **8.** 0   **9.** 1   **10.** $-\dfrac{1}{2}$

11. $\dfrac{\sqrt{11}}{11}$ **12.** $\dfrac{\sqrt{6}}{3}$ **13.** IV   **14.** II

15. II   **16.** impossible   **17.** possible   **18.** possible

19. $\dfrac{3}{4}$

20. $-\dfrac{\sqrt{2}}{2}$

21. $\dfrac{2\sqrt{5}}{5}$

22. $\dfrac{15}{17}$; $-\dfrac{8}{17}$; $-\dfrac{15}{8}$; $-\dfrac{8}{15}$; $-\dfrac{17}{8}$; $\dfrac{17}{15}$

23. $-\dfrac{12}{13}$; $\dfrac{5}{13}$; $-\dfrac{12}{5}$; $-\dfrac{5}{12}$; $\dfrac{13}{5}$; $-\dfrac{13}{12}$

5.3 Evaluating Trigonometric Functions

In exercises 7-9, we give, in order sine, cosine, tangent, cotangent, secant, and cosecant.

1. $\dfrac{\sqrt{3}}{3}$

2. $\dfrac{\sqrt{3}}{2}$

3. 2

4. 53°

5. 38°

6. 64°

7. $\dfrac{\sqrt{2}}{2}$; $-\dfrac{\sqrt{2}}{2}$; -1; -1; $-\sqrt{2}$; $\sqrt{2}$

8. $\dfrac{1}{2}$; $\dfrac{\sqrt{3}}{2}$; $\dfrac{\sqrt{3}}{3}$; $\sqrt{3}$; $\dfrac{2\sqrt{3}}{3}$; 2

9. $\dfrac{1}{2}$; $\dfrac{\sqrt{3}}{2}$; $\dfrac{\sqrt{3}}{3}$; $\sqrt{3}$; $\dfrac{2\sqrt{3}}{3}$; 2

10. $\dfrac{\sqrt{3}}{2}$

11. $-\dfrac{\sqrt{3}}{2}$

12. $\dfrac{\sqrt{3}}{3}$

13. True

14. False

15. False

16. -1.88708

17. -1.04135

18. -0.20709

19. 49.996°

20. 72.001°

21. 14.8101°

22. 240°, 300°

23. 60°, 300°

24. 45°, 225°

5.4 Solving Right Triangles

1. $B = 60°$, $a = 57.7$, $c = 115.5$
2. $A = 55°$, $a = 7.1$, $c = 8.7$
3. $B = 57.4°$, $b = 65$, $c = 77.2$
4. $B = 40°$, $a = 9.2$, $b = 7.7$
5. $A = 50°$, $a = 7$, $b = 5.9$
6. $A = 57.46°$, $b = 4.2$, $c = 7.9$
7. $A = 38°$, $b = 39.7$, $c = 50.4$
8. $B = 15°$, $a = 19.3$, $b = 5.2$
9. $B = 22°$, $a = 198.1$, $c = 213.6$
10. $A = 66.2°$, $B = 23.8°$, $a = 47.79$
11. $A = 60°$, $b = 23.1$, $c = 46.2$
12. $A = 36.9°$, $B = 53.1°$, $c = 25$
13. $A = 52°$, $b = 11.7$, $c = 19$
14. $A = 56.2°$, $B = 33.8°$, $b = 3.661$
15. $A = 31°$, $B = 59°$, $a = 72.1$
16. $A = 33.6°$, $B = 56.4°$, $b = 42.7$
17. $B = 30°$, $a = 19.1$, $b = 10.9$
18. $A = 64.6°$, $a = 2.9$, $b = 1.4$
19. $B = 12°$, $a = 940.9$, $c = 961.9$
20. $A = 20°$, $B = 70°$, $c = 87.71$

Convert each degree measure to radians. Leave answers as multiples of π.

1. 15°
2. -300°
3. 720°

Convert each degree measure to radians.

4. 328°
5. 4.8°
6. π°

Convert each radian measure to degrees.

7. $\dfrac{11\pi}{18}$
8. $\dfrac{7\pi}{3}$
9. $-\dfrac{\pi}{80}$
10. 5.9
11. -0.5
12. 2.6

Find the length to three significant digits of each arc intercepted by a central angle θ in a circle of radius r.

13. $r = 4$ cm; $\theta = \dfrac{\pi}{8}$
14. $r = 39$ in; $\theta = 25°$
15. $r = 13$ m; $\theta = 21°$
16. $r = 3$ cm; $\theta = 1$
17. $r = 15$ m; $\theta = 76°$
18. $r = 80$ in; $\theta = \dfrac{3\pi}{2}$

Find the area of a sector of a circle having radius r and central angle θ. Express answers to the nearest tenth

19. $r = 1.4$ ft; $\theta = \dfrac{5\pi}{4}$
20. $r = 5$ in; $\theta = 2$
21. $r = 4$ cm; $\theta = 3\pi$
22. $r = 15$ ft; $\theta = 31°$
23. $r = 9.5$ in; $\theta = 24°$
24. $r = 9$ cm; $\theta = 60°$

6.2 The Unit Circle and Circular Functions

Find the exact circular function value for each of the following.

1. $\sin\dfrac{5\pi}{6}$

2. $\cos\dfrac{11\pi}{6}$

3. $\tan\dfrac{13\pi}{6}$

4. $\csc 12\pi$

5. $\sec\left(-\dfrac{8\pi}{3}\right)$

Find a calculator approximation for each circular function value.

6. $\sec(1.324)$

7. $\cos(-4)$

8. $\sin(2.268)$

9. $\csc(2.805)$

10. $\cot(-2.512)$

Find the exact value of s in the given interval that has the given circular function value.

11. $[-\dfrac{\pi}{2}, -\pi]$; $\sin s = -\dfrac{1}{2}$

12. $[\pi, \dfrac{3\pi}{2}]$; $\tan s = 1$

13. $[\dfrac{\pi}{2}, \pi]$; $\csc s = 2$

14. $[\dfrac{3\pi}{2}, 2\pi]$; $\cos s = \dfrac{1}{2}$

15. $[\dfrac{\pi}{2}, \pi]$; $\sec s = -\dfrac{2\sqrt{3}}{3}$

Use the formula $\omega = \theta / t$ to find the value of the missing variable.

16. $\theta = \dfrac{3\pi}{4}$; $t = 2$ min

17. $\omega = 5$ rad/sec; $t = 3$ sec

18. $\theta = 12\pi$; $t = 3.6$ hr

19. $\omega = 4$ rad/sec; $t = 10$ sec

20. $\theta = 12$; $t = 3$ min

21. $\omega = \dfrac{4\pi}{3}$ rad/sec; $t = 30$ sec

22. $\theta = 15\pi$; $t = 2.4$ hr

Use the formula $v = r \cdot \omega$ to find the value of the missing variable.

23. $\omega = \dfrac{\pi}{12}$ rad/sec; $r = 936$ ft

24. $\omega = 8$ rad/sec; $r = 4$ in

25. $r = 3$ cm; $v = 6$ cm/sec

6.3 Graphs of the Sine and Cosine Functions

Find the amplitude of the graph of each function. Then give a calculator window that shows the function over a two-period interval.

1. $y = 5\cos x$

2. $y = -4\sin x$

3. $y = 7\cos x$

4. $y = 2\sin x$

5. $y = -4\cos x$

6. $y = \dfrac{5}{8}\sin x$

7. $y = 8\cos x$

Find the amplitude and the period of the graph of each function. Then give a calculator window that shows the function over a two-period interval.

8. $y = -\dfrac{1}{2}\cos(\pi x)$

9. $y = \sin(2x)$

10. $y = \dfrac{1}{4}\sin(\pi x)$

11. $y = -3\cos\left(\dfrac{x}{4}\right)$

12. $y = 3\sin(4\pi x)$

13. $y = -2\sin(6x)$

14. $y = -3\sin\left(\dfrac{x}{2}\right)$

15. $y = -\dfrac{4}{3}\cos(3x)$

16. $y = \dfrac{1}{2}\sin\left(\dfrac{\pi x}{3}\right)$

17. $y = \cos\left(\dfrac{1}{3}x\right)$

18. $y = 4\sin(6x)$

19. $y = 3\cos\left(\dfrac{x}{2}\right)$

20. $y = -\dfrac{4}{3}\sin\left(\dfrac{x}{\pi}\right)$

6.4 Translations of the Graphs of the Sine and Cosine Functions

Find the amplitude, the period, any vertical translation and any phase shift of the graph of each function. Then give a calculator window that shows the function over a two-period interval.

1. $y = \cos\left(x - \dfrac{\pi}{2}\right)$

2. $y = \cos\left(x - \dfrac{\pi}{4}\right)$

3. $y = 4\cos\left(x + \dfrac{\pi}{4}\right) + 5$

4. $y = \dfrac{1}{2}\sin\left(\dfrac{\pi x}{2}\right)$

5. $y = -2 + 4\cos(2x)$

6. $y = -2 + 2\cos\left(\dfrac{x}{2} + \dfrac{\pi}{2}\right)$

7. $y = 3 + 3\sin\left(\dfrac{x}{2} + \dfrac{\pi}{3}\right)$

8. $y = -4\cos\left(\dfrac{2x}{3} + \dfrac{\pi}{6}\right)$

9. $y = 6\sin\left(\dfrac{x}{3} - \dfrac{\pi}{6}\right)$

10. $y = 4\sin(\pi x - 2) + 1$

11. $y = \sin\left(x + \dfrac{\pi}{6}\right)$

12. $y = 2\sin\left(3x - \dfrac{\pi}{2}\right)$

13. $y = \cos(2x - \pi)$

14. $y = 3\sin(2\pi x) - 2$

15. $y = -\sin\left(x - \dfrac{\pi}{2}\right)$

16. $y = 2\cos\left(x - \dfrac{5\pi}{6}\right)$

17. $y = \dfrac{1}{2}\sin\left(x + \dfrac{\pi}{4}\right)$

18. $y = 4\sin\left(\dfrac{x}{2} + \dfrac{\pi}{2}\right)$

19. $y = 2\sin\left(\dfrac{x}{3}\right) - 1$

20. $y = \cos\left(\dfrac{x}{3} - \dfrac{\pi}{2}\right) - 1$

6.5 Graphs of the Tangent, Cotangent, Secant, and Cosecant Functions

Graph each function over a two-period interval. Give a calculator window that shows the graph.

1. $y = 3\tan\left(\dfrac{x}{3}\right)$

2. $y = \tan\left(x - \dfrac{\pi}{4}\right)$

3. $y = -2\csc(3x)$

4. $y = 2\cot\left(2x - \dfrac{\pi}{4}\right)$

5. $y = -\cot\left(\dfrac{x}{\pi}\right)$

6. $y = -\sec(4x - \pi)$

7. $y = \sec\left(x + \dfrac{\pi}{4}\right)$

8. $y = \csc(2x + \pi)$

9. $y = -3\cot\left(\dfrac{x}{4} + 2\pi\right)$

10. $y = \sec(3x)$

11. $y = \csc(\pi x) - 1$

12. $y = 3\csc\left(2x - \dfrac{\pi}{3}\right)$

13. $y = \tan\left(\dfrac{x}{2}\right) - 4$

14. $y = 3\csc\left(2x + \dfrac{\pi}{3}\right)$

15. $y = \csc\left(x + \dfrac{\pi}{4}\right)$

16. $y = 3\tan\left(\dfrac{\pi x}{2}\right)$

17. $y = \cot\left(2x + \dfrac{\pi}{2}\right)$

18. $y = \sec\left(\dfrac{x}{2}\right) - 1$

19. $y = -2\csc(4x) + 2$

20. $y = -2 + \cot\left(\dfrac{x}{3} - \dfrac{\pi}{2}\right)$

6.6 Harmonic Motion

Suppose that a weight on a spring has initial position $s(0)$ and period P. Find a function s given by $s(t) = a\cos(\omega t)$ that models the displacement of the weight.

1. $s(0) = 3$ inches; $P = 0.6$ second
2. $s(0) = 6$ inches; $P = 2.3$ seconds
3. $s(0) = -2$ inches; $P = 0.4$ second
4. $s(0) = -5$ inches; $P = 1.5$ seconds

A note on the piano has given frequency F. Suppose the maximum displacement at the center of the piano wire is given by $s(0)$. Find constants a and ω so that the equation $s(t) = a\cos(\omega t)$ models this displacement.

5. $F = 32$; $s(0) = 0.17$
6. $F = 140$; $s(0) = 0.07$
7. $F = 66$; $s(0) = 0.24$
8. $F = 280$; $s(0) = 0.12$

The period P, in seconds, of oscillation of a pendulum of length L feet is given by $P = 2\pi\sqrt{\dfrac{L}{32}}$. Find the missing value.

9. $L = 1$ foot
10. $L = 2.5$ feet
11. $P = 6$ seconds
12. $P = 0.5$ second

Solve each problem.

13. The equation for the simple harmonic motion of a particle is given by $s(t) = 3\cos\left(\dfrac{3\pi}{2}t\right)$. Find the maximum displacement and the frequency.

14. The position of a weight attached to a spring is $s(t) = -8\cos(16\pi t)$ inches after t seconds. What is the maximum height that the weight rises above the equilibrium position and when does the weight first reach its maximum height?

15. The position of a weight attached to a spring is $s(t) = -\dfrac{1}{2}\cos(20\pi t)$ inches after t seconds. What are the frequency and period?

16. The position of a weight attached to a spring is $s(t) = -8\cos(60\pi t)$ inches after t seconds. Calculate and interpret $s(0.26)$.

17. A weight attached to a spring is pulled down 6 inches below the equilibrium position. Assuming that the frequency is $7/\pi$ cycles per second, determine a trigonometric model that gives the position of the weight at time t seconds.

Answers

6.1 Radian Measure

1. $\dfrac{\pi}{12}$
2. $-\dfrac{5\pi}{3}$
3. 4π
4. 5.72
5. 0.08
6. 0.05
7. 110°
8. 420°
9. -2.25°
10. 338.05°
11. -28.65°
12. 148.97°
13. 1.57 cm
14. 17.0 in
15. 4.76 m
16. 3 cm
17. 19.9 m
18. 377 in
19. 3.8 ft^2
20. 25 in^2
21. 75.4 cm^2
22. 60.9 ft^2
23. 18.9 in^2
24. 42.4 cm^2

6.2 The Unit Circle and Circular Functions

1. $\dfrac{1}{2}$
2. $\dfrac{\sqrt{3}}{2}$
3. $\dfrac{\sqrt{3}}{3}$
4. undefined
5. –2
6. 4.093
7. –0.6536
8. 0.7666
9. 3.0278
10. 1.3727
11. $-\dfrac{5\pi}{6}$
12. $\dfrac{5\pi}{4}$
13. $\dfrac{5\pi}{6}$
14. $\dfrac{5\pi}{3}$
15. $\dfrac{5\pi}{6}$
16. $\dfrac{3\pi}{8}$ rad/min
17. 15 rad
18. 10.47 rad/hr
19. 40 rad
20. 4 rad/min
21. 40π rad
22. 19.64 rad/hr
23. 78π ft/s
24. 32 in/s
25. 2 rad/sec

6.3 Graphs of the Sine and Cosine Functions

1. 5; $[-2\pi, 2\pi]$ by $[-6,6]$

2. 4; $[-2\pi, 2\pi]$ by $[-5,5]$

3. 7; $[-2\pi, 2\pi]$ by $[-8,8]$

4. 2; $[-2\pi, 2\pi]$ by $[-3,3]$

5. 4; $[-2\pi, 2\pi]$ by $[-5,5]$

6. $\frac{5}{8}$; $[-2\pi, 2\pi]$ by $[-1,1]$

7. 8; $[-2\pi, 2\pi]$ by $[-9,9]$

8. $\frac{1}{2}$; 2; $[-2,2]$ by $[-1,1]$

9. 1; π; $[-\pi, \pi]$ by $[-2,2]$

10. $\frac{1}{4}$; 2; $[-2,2]$ by $[-1,1]$

11. 3; 8π; $[-8\pi, 8\pi]$ by $[-4,4]$

12. 3; $\frac{1}{2}$; $[-\frac{1}{2}, \frac{1}{2}]$ by $[-4,4]$

13. 2; $\frac{\pi}{3}$; $[-\frac{\pi}{3}, \frac{\pi}{3}]$ by $[-3,3]$

14. 3; 4π; $[-4\pi, 4\pi]$ by $[-4,4]$

15. $-\frac{4}{3}$; $\frac{2\pi}{3}$; $[-\frac{2\pi}{3}, \frac{2\pi}{3}]$ by $[-2,2]$

16. $\frac{1}{2}$; 6; $[-6,6]$ by $[-1,1]$

17. 1; 6π; $[-6\pi, 6\pi]$ by $[-2,2]$

18. 4; $\frac{\pi}{3}$; $[-\frac{\pi}{3}, \frac{\pi}{3}]$ by $[-5,5]$

19. 3; 4π; $[-4\pi, 4\pi]$ by $[-4,4]$

20. $\frac{4}{3}$; $2\pi^2$; $[-2\pi^2, 2\pi^2]$ by $[-2,2]$

6.4 Translations of the Graphs of the Sine and Cosine Functions

1. 1; 2π; none; $\frac{\pi}{2}$ right; $[-\frac{3\pi}{2}, \frac{5\pi}{2}]$ by $[-2,2]$

2. 1; 2π; none; $\frac{\pi}{4}$ right; $[-\frac{7\pi}{4}, \frac{9\pi}{4}]$ by $[-2,2]$

3. 4; 2π; 5 up; $\frac{\pi}{4}$ left; $[-\frac{9\pi}{4}, \frac{7\pi}{4}]$ by $[0,10]$

4. $\frac{1}{2}$; 4; none; none; $[-4,4]$ by $[-1,1]$

5. 4; π; 2 down; none; $[-\pi, \pi]$ by $[-7,3]$

6. 2; 4π; 2 down; π left; $[-5\pi, 3\pi]$ by $[-4,0]$

Answers

7. 3; 4π; 3 up; $\frac{2\pi}{3}$ left; $[-\frac{14\pi}{3}, \frac{10\pi}{3}]$ by $[0,6]$

8. 4; 3π; none; $\frac{\pi}{4}$ left; $[-\frac{13\pi}{4}, \frac{11\pi}{4}]$ by $[-5,5]$

9. 6; 6π; none; $\frac{\pi}{2}$ right; $[-\frac{11\pi}{2}, \frac{13\pi}{2}]$ by $[-7,7]$

10. 4; 2; 1 up; $\frac{2}{\pi}$ right; $[-2+\frac{2}{\pi}, 2+\frac{2}{\pi}]$ by $[-4,6]$

11. 1; 2π; none; $\frac{\pi}{6}$ left; $[-\frac{13\pi}{6}, \frac{11\pi}{6}]$ by $[-2,2]$

12. 2; $\frac{2\pi}{3}$; none; $\frac{\pi}{6}$ right; $[-\frac{\pi}{2}, \frac{5\pi}{6}]$ by $[-3,3]$

13. 1; π; none; $\frac{\pi}{2}$ right; $[-\frac{\pi}{2}, \frac{3\pi}{2}]$ by $[-2,2]$

14. 3; 1; 2 down; none; $[-1,1]$ by $[-6,2]$

15. 1; 2π; none; $\frac{\pi}{2}$ right; $[-\frac{3\pi}{2}, \frac{5\pi}{2}]$ by $[-2,2]$

16. 2; 2π; none; $\frac{5\pi}{6}$ right; $[-\frac{7\pi}{6}, \frac{17\pi}{6}]$ by $[-3,3]$

17. $\frac{1}{2}$; 2π; none; $\frac{\pi}{4}$ left; $[-\frac{9\pi}{4}, \frac{7\pi}{4}]$ by $[-1,1]$

18. 4; 4π; none; π left; $[-5\pi, 3\pi]$ by $[-5,5]$

19. 2; 6π; 1 down; none; $[-6\pi, 6\pi]$ by $[-4,2]$

20. 1; 6π; 1 down; $\frac{3\pi}{2}$ right; $[-\frac{9\pi}{2}, \frac{15\pi}{2}]$ by $[-3,1]$

6.5 Graphs of the Tangent, Cotangent, Secant, and Cosecant Functions

1. $[-3\pi, 3\pi]$ by $[-4,4]$

2. $[-\frac{3\pi}{4}, \frac{5\pi}{4}]$ by $[-4,4]$

3. $[-\frac{2\pi}{3}, \frac{2\pi}{3}]$ by $[-4,4]$

4. $[-\frac{3\pi}{8}, \frac{5\pi}{8}]$ by $[-4,4]$

5. $[-\pi^2, \pi^2]$ by $[-4,4]$

6. $[-\frac{\pi}{4}, \frac{3\pi}{4}]$ by $[-4,4]$

7. $[-\frac{9\pi}{4}, \frac{7\pi}{4}]$ by $[-4,4]$

8. $[-\frac{3\pi}{2}, \frac{\pi}{2}]$ by $[-4,4]$

9. $[-8\pi, 0]$ by $[-4,4]$

10. $[-\frac{2\pi}{3}, \frac{2\pi}{3}]$ by $[-4,4]$

11. $[-2,2]$ by $[-4,4]$

12. $[-\frac{5\pi}{6}, \frac{7\pi}{6}]$ by $[-6,6]$

13. $[-2\pi, 2\pi]$ by $[-8,4]$

14. $[-\frac{7\pi}{6}, \frac{5\pi}{6}]$ by $[-6,6]$

15. $[-\frac{9\pi}{4}, \frac{7\pi}{4}]$ by $[-4,4]$

16. $[-2,2]$ by $[-4,4]$

17. $[-\frac{3\pi}{4}, \frac{\pi}{4}]$ by $[-4,4]$

18. $[-4\pi, 4\pi]$ by $[-4,4]$

19. $[-\frac{\pi}{2}, \frac{\pi}{2}]$ by $[-4,8]$

20. $[-\frac{3\pi}{2}, \frac{9\pi}{2}]$ by $[-6,4]$

6.6 Harmonic Motion

1. $s(t) = 3\cos\left(\dfrac{10\pi}{3}t\right)$
2. $s(t) = 6\cos\left(\dfrac{20\pi}{23}t\right)$
3. $s(t) = -2\cos(5\pi\, t)$
4. $s(t) = -5\cos\left(\dfrac{4\pi}{3}t\right)$
5. $s(t) = 0.17\cos(64\pi\, t)$
6. $s(t) = 0.07\cos(280\pi\, t)$
7. $s(t) = 0.24\cos(132\pi\, t)$
8. $s(t) = 0.12\cos(560\pi\, t)$
9. $\sqrt{2}\pi/4$ sec
10. $\sqrt{5}\pi/4$ sec
11. $288/\pi^2$ feet
12. $2/\pi^2$ feet
13. 3; 3/4
14. 8; 1/16 sec
15. 10 cycles/sec; 1/10 sec
16. -2.47 ; After 0.26 seconds the weight is about 2.5 inches below the equilibrium position.
17. $s(t) = -6\cos(14t)$

7.1 Fundamental Identities

Find $\sin\theta$.

1. $\cos\theta = -\dfrac{12}{13}$, θ in quadrant III

2. $\tan\theta = \dfrac{9}{40}$, θ in quadrant I

3. $\sec\theta = -\dfrac{4}{\sqrt{5}}$, $\tan\theta < 0$

Find the remaining five trigonometric functions of θ.

4. $\sin\theta = -\dfrac{8}{17}$, $\cos\theta < 0$

5. $\tan\theta = \dfrac{1}{2}$, θ in quadrant I

6. $\sec\theta = 3$, θ in quadrant IV

7. $\sin\theta = \dfrac{1}{3}$, θ in quadrant II

8. $\csc\theta = \sqrt{3}$, $\cos\theta > 0$

Write each expression in terms of sine and cosine, and simplify so that no quotients appear in the final expression.

9. $\tan^2\theta \cdot \csc\theta \cdot \cos\theta$

10. $\sin\theta \cdot \cos\theta \cdot \csc\theta$

11. $(\sin\theta + \cos\theta)^2 - 2\sin\theta \cdot \cos\theta$

12. $\tan^2\theta - \cot^2\theta$

13. $\sin^2\theta + \cot^2\theta \cdot \sin^2\theta$

14. $(1+\sin\theta)(1-\sin\theta)$

15. $\dfrac{1}{\sin^2\theta} - \dfrac{1}{\tan^2\theta}$

16. $\tan\theta(\cos\theta - \csc\theta)$

17. $\sin\theta(\sec\theta + \cot\theta)$

Use a graphing calculator to make a conjecture as to whether each equation is an identity.

18. $\tan^2 x = \sec^2 x - 1$

19. $\cos^2 x = 2 + \cos 2x$

20. $\tan 2x = 1 - \tan^2 x$

7.2 Verifying Trigonometric Identities

Perform each indicated operation and simplify the result.

1. $\dfrac{1}{\sec x + 1} - \dfrac{1}{\sec x - 1}$

2. $(1 - \csc x)(1 + \csc x)$

3. $\dfrac{\cos x + \sin x}{\cos x}$

4. $\cos x(\csc x - \sec x)$

Factor each trigonometric expression.

5. $\tan^2 \theta - \tan^2 \theta \sin^2 \theta$

6. $\sec^2 x - 4$

7. $\cos^4 \theta - \sin^4 \theta$

Use fundamental identities to simplify each expression.

8. $\tan \theta \csc \theta$

9. $\sin^2 \theta (\cot^2 \theta + 1)$

10. $\dfrac{\csc \theta}{\sec \theta}$

Verify that each trigonometric equation is an identity.

11. $\dfrac{\cot^3 \theta}{\csc \theta} = \cos \theta (\csc^2 \theta - 1)$

12. $\dfrac{\tan x \cdot \cos x}{\sin x} = 1$

13. $\sec x - \cos x = \tan x \cdot \sin x$

14. $\dfrac{\sin x + \cos x}{\sin x} = 1 + \cot x$

15. $\dfrac{\sin x - \cos x}{\sin x \cdot \cos x} = \sec x - \csc x$

16. $\dfrac{1}{\sec x \cdot \tan x} = \csc x - \sin x$

17. $\tan x (1 - \cot x) = \tan x - 1$

18. $(\sin x + \cos x)^2 + (\sin x - \cos x)^2 = 2$

19. $\sec^2 x + \sec x \cdot \tan x = \dfrac{1}{1 - \sin x}$

20. $\dfrac{1 + \cos x}{\sin x} = \dfrac{1}{\csc x - \cot x}$

7.3 Sum and Difference Identities

Use identities to find each exact value.

1. cos 165°

2. cos 255°

3. $\cos\left(\dfrac{11\pi}{12}\right)$

Write each function value in terms of the cofunction of a complementary angle.

4. sin 75°

5. cos 34°

6. $\tan\left(\dfrac{\pi}{10}\right)$

Find an angle θ that makes each statement true.

7. $\sin\theta = \cos(\theta + 10°)$

8. $\tan(2\theta - 14°) = \cot(\theta + 20°)$

Use identities to find the exact value of each of the following.

9. cos 75°

10. $\cos\left(\dfrac{\pi}{12}\right)$

11. $\sin\left(\dfrac{21\pi}{12}\right)$

12. $\cos\left(\dfrac{10\pi}{9}\right)\cdot\cos\left(\dfrac{5\pi}{36}\right) - \sin\left(\dfrac{10\pi}{9}\right)\cdot\sin\left(\dfrac{5\pi}{36}\right)$

Use identities to write each expression as a function of x or θ.

13. $\sin\left(\dfrac{3\pi}{2} + \theta\right)$

14. $\tan\left(\dfrac{\pi}{4} - x\right)$

15. $\cos(x + 180°)$

16. $\cos(2\pi - \theta)$

Use the given information to find $\cos(s+t)$ and $\tan(s-t)$.

17. $\sin s = \dfrac{5}{13}$; $\cos t = -\dfrac{3}{5}$; s in quadrant I, t in quadrant II

18. $\sin s = \dfrac{3}{5}$; $\cos t = -\dfrac{5}{13}$; s in quadrant I, t in quadrant II

Verify that each equation is an identity.

19. $\sin(x+y)\cdot\sin(x-y) = \sin^2 x - \sin^2 y$

7.4 Double-Angle and Half-Angle Identities

Use identities to find values of the sine and cosine functions for each angle measure.

1. θ, given $\cos 2\theta = \dfrac{5}{13}$ and $0 < \theta < \dfrac{\pi}{2}$

2. θ, given $\cos 2\theta = \dfrac{1}{3}$ and $\pi < \theta < \dfrac{3\pi}{2}$

3. 2θ, given $\tan \theta = \dfrac{5}{12}$ and $\cos \theta > 0$

4. 2θ, given $\sin \theta = \dfrac{24}{25}$ and $\tan \theta < 0$

Use an identity to write each expression as a single trigonometric function value or as a single number.

5. $2 \sin 15° \cos 15°$

6. $\dfrac{2 \tan 75°}{1 - \tan^2 75°}$

7. $\cos^2 \dfrac{\pi}{8} - \sin^2 \dfrac{\pi}{8}$

8. $1 - 2\sin^2\left(-\dfrac{\pi}{8}\right)$

Verify that each equation is an identity.

9. $\tan \dfrac{\theta}{2} = \csc \theta - \cot \theta$

10. $\cos 8\theta = \cos^2 4\theta - \sin^2 4\theta$

11. $(\sin \theta + \cos \theta)^2 = 1 + \sin 2\theta$

12. $\dfrac{2 \tan \theta}{1 + \tan^2 \theta} = \sin 2\theta$

13. $\dfrac{\cos 2\theta}{\sin^2 \theta} = \csc^2 \theta - 2$

14. $\sin^2 \theta = \dfrac{1 - \cos 2\theta}{2}$

Find each of the following.

15. $\sin \dfrac{\theta}{2}$, given $\cos \theta = \dfrac{1}{2}$ with $\dfrac{3\pi}{2} < \theta < 2\pi$

16. $\cos \dfrac{\theta}{2}$, given $\sin \theta = \dfrac{3}{5}$ with $\dfrac{\pi}{2} < \theta < \pi$

17. $\tan \dfrac{\theta}{2}$, given $\sec \theta = -\dfrac{41}{40}$ with $\dfrac{\pi}{2} < \theta < \pi$

Use an identity to write each expression with a single trigonometric function.

18. $\sqrt{\dfrac{1 - \cos 30°}{2}}$

19. $\sqrt{\dfrac{1 + \cos 80°}{2}}$

20. $\dfrac{\sin 50°}{1 + \cos 50°}$

7.5 Inverse Circular Functions

Find the exact value of each real number y.

1. $\sin^{-1}\left(\dfrac{1}{\sqrt{2}}\right)$

2. $\arcsin(1)$

3. $\tan^{-1}\left(-\sqrt{3}\right)$

4. $\cos^{-1}\left(\dfrac{\sqrt{3}}{2}\right)$

5. $\cos^{-1}\left(-\dfrac{\sqrt{2}}{2}\right)$

6. $\arctan\left(-\dfrac{1}{\sqrt{3}}\right)$

7. $\csc^{-1}(1)$

Give the degree measure of θ.

8. $\theta = \sin^{-1}\left(\dfrac{1}{2}\right)$

9. $\theta = \arccos\left(-\dfrac{\sqrt{3}}{2}\right)$

10. $\theta = \tan^{-1}(1)$

Use a calculator to give each value in decimal degrees.

11. $\theta = \arcsin(0.1111111)$

12. $\theta = \cos^{-1}(-0.92170128)$

Give the exact value of each expression.

13. $\cos(\tan^{-1} 2)$

14. $\sec\left(\tan^{-1}\dfrac{12}{5}\right)$

15. $\tan\left(\cos^{-1}\dfrac{3}{5}\right)$

16. $\sec^{-1}\left(\tan\dfrac{3\pi}{4}\right)$

Write each expression as an algebraic expression in u, $u > 0$.

17. $\csc\left(\sin^{-1}\dfrac{1}{u}\right)$

18. $\cos(2\cos^{-1} u)$

7.6 Trigonometric Equations

Solve each equation for exact solutions in the interval $[0, 2\pi)$.

1. $2\cos x + 1 = 0$
2. $2\sin^2 x - 3\sin x + 1 = 0$
3. $\sqrt{2}\cos x + 1 = 0$
4. $4\csc^2 x + 3\csc x - 1 = 0$

Solve each equation for exact solutions in the interval $[0°, 360°)$.

5. $(\sin\theta - 1)(2\sin\theta - 1) = 0$
6. $\sec\theta \cdot \csc\theta - 2\csc\theta = 0$
7. $\sin^2\theta = 1$
8. $\tan^2\theta = \tan\theta$
9. $2\sin^2\theta - \sin\theta = 1$
10. $4\sin^3\theta - 3\sin\theta = 0$

Determine all solutions of each equation in radians.

11. $\cos^3 x = \cos x$
12. $2\sin^2 x \cdot \cos x - \cos x = 0$

Solve each equation for exact solutions in the interval $[0, 2\pi)$.

13. $\sin 2x = \dfrac{\sqrt{3}}{2}$
14. $\cos^2 4x = 1$
15. $\sin 2x = \sqrt{3}\sin x$
16. $\tan 3x = 1$

Solve each equation for exact solutions in the interval $[0°, 360°)$.

17. $\cos^2 2\theta - 2\sin 2\theta + 2 = 0$
18. $\cos 2\theta \cdot \cos\theta - \sin 2\theta \cdot \sin\theta = \dfrac{1}{2}$
19. $\tan\dfrac{\theta}{2} = \sin\theta$
20. $\cos\theta = \sin\dfrac{\theta}{2}$

7.7 Equations Involving Inverse Trigonometric Functions

Solve each equation for x.

1. $y = \sec(2x) + \sqrt{5}$

2. $y = 10\sin x$

3. $4y = \tan(7x)$

4. $y = 5\cos(4x)$

5. $y = 5\sin\dfrac{x}{3}$

6. $y = -7\csc(3x)$

7. $y = \sin(x-1)$

8. $y = \tan x + 5$

9. $y = 3\cos x - 5$

10. $y = 3 - 7\tan x$

Solve each equation for exact solutions.

11. $6\sin^{-1}(\dfrac{x}{8}) = \pi$

12. $2\sin^{-1}(\dfrac{\pi+6}{x}) = \pi$

13. $\tan^{-1}(\dfrac{12}{5}) = \cos^{-1} x$

14. $\sin^{-1}(\dfrac{2}{\sqrt{5}}) = \tan^{-1} x$

15. $\cos^{-1}\left(\dfrac{3}{\sqrt{10}}\right) = \sin^{-1} x$

Solve each equation for exact solutions.

16. $\sin^{-1}\dfrac{1}{2} + \cos^{-1} x = \pi$

17. $\sin^{-1} x - \tan^{-1}\sqrt{3} = \dfrac{\pi}{6}$

18. $\cos^{-1} x + \tan^{-1} x = \dfrac{3\pi}{4}$

19. $\sin^{-1} x - \cos^{-1} 2x = \dfrac{\pi}{6}$

20. $\sin^{-1}(-\dfrac{1}{2}) + 2\tan^{-1} x = \dfrac{\pi}{2}$

Answers

7.1 Fundamental Identities

1. $-\dfrac{5}{13}$
2. $\dfrac{9}{41}$
3. $\dfrac{\sqrt{11}}{4}$

4. $\cos\theta = -\dfrac{15}{17}$, $\tan\theta = \dfrac{8}{15}$, $\csc\theta = -\dfrac{17}{8}$, $\sec\theta = -\dfrac{17}{15}$, $\cot\theta = \dfrac{15}{8}$

5. $\sin\theta = \dfrac{\sqrt{5}}{5}$, $\cos\theta = \dfrac{2\sqrt{5}}{5}$, $\csc\theta = \sqrt{5}$, $\sec\theta = \dfrac{\sqrt{5}}{2}$, $\cot\theta = 2$

6. $\sin\theta = -\dfrac{2\sqrt{2}}{3}$, $\cos\theta = \dfrac{1}{3}$, $\tan\theta = -2\sqrt{2}$, $\csc\theta = -\dfrac{3\sqrt{2}}{4}$, $\cot\theta = -\dfrac{\sqrt{2}}{4}$

7. $\cos\theta = -\dfrac{2\sqrt{2}}{3}$, $\tan\theta = -\dfrac{\sqrt{2}}{4}$, $\csc\theta = 3$, $\sec\theta = -\dfrac{3\sqrt{2}}{4}$, $\cot\theta = -2\sqrt{2}$

8. $\sin\theta = \dfrac{\sqrt{3}}{3}$, $\cos\theta = \dfrac{\sqrt{6}}{3}$, $\tan\theta = \dfrac{\sqrt{2}}{2}$, $\sec\theta = \dfrac{\sqrt{6}}{2}$, $\cot\theta = \sqrt{2}$

9. $\tan\theta$
10. $\cos\theta$
11. 1
12. $\sec^2\theta - \csc^2\theta$
13. 1
14. $\cos^2\theta$
15. 1
16. $\sin\theta - \sec\theta$
17. $\tan\theta + \cos\theta$
18. identity
19. not an identity
20. not an identity

7.2 Verifying Trigonometric Identities

1. $-2\cot^2 x$
2. $-\cot^2 x$
3. $1 + \tan x$
4. $\cot x - 1$
5. $\sin^2\theta$
6. $(\sec x + 2)(\sec x - 2)$
7. $(\cos\theta - \sin\theta)(\cos\theta + \sin\theta)$
8. $\sec\theta$
9. 1
10. $\cot\theta$

7.3 Sum and Difference Identities

1. $\dfrac{-\sqrt{6}-\sqrt{2}}{4}$
2. $\dfrac{-\sqrt{6}+\sqrt{2}}{4}$
3. $\dfrac{-\sqrt{6}-\sqrt{2}}{4}$
4. $\cos 15°$
5. $\sin 56°$
6. $\cot \dfrac{2\pi}{5}$
7. $40°$
8. $28°$
9. $\dfrac{\sqrt{6}-\sqrt{2}}{4}$
10. $\dfrac{\sqrt{6}+\sqrt{2}}{4}$
11. $-\dfrac{\sqrt{2}}{2}$
12. $-\dfrac{\sqrt{2}}{2}$
13. $-\cos\theta$
14. $\dfrac{1-\tan x}{1+\tan x}$
15. $-\cos x$
16. $\cos\theta$
17. $-\dfrac{56}{65},\dfrac{63}{16}$
18. $-\dfrac{56}{65},-\dfrac{63}{16}$

7.4 Double-Angle Identities and Half-Angle Identities

1. $\dfrac{2\sqrt{13}}{13};\dfrac{3\sqrt{13}}{13}$
2. $-\dfrac{\sqrt{3}}{3};-\dfrac{\sqrt{6}}{3}$
3. $\dfrac{120}{169};\dfrac{119}{169}$
4. $-\dfrac{336}{625};-\dfrac{527}{625}$
5. $\dfrac{1}{2}$
6. $-\dfrac{\sqrt{3}}{3}$
7. $\dfrac{\sqrt{2}}{2}$
8. $\dfrac{\sqrt{2}}{2}$
15. $\dfrac{1}{2}$
16. $\dfrac{\sqrt{10}}{10}$
17. 9
18. $\sin 15°$
19. $\cos 40°$
20. $\tan 25°$

7.5 Inverse Circular Functions

1. $\dfrac{\pi}{4}$
2. $\dfrac{\pi}{2}$
3. $-\dfrac{\pi}{3}$
4. $\dfrac{\pi}{6}$

5. $\dfrac{3\pi}{4}$
6. $-\dfrac{\pi}{6}$
7. $\dfrac{\pi}{2}$
8. 30°

9. 150°
10. 45°
11. 6.37937°
12. 157.176°

13. $\dfrac{\sqrt{5}}{5}$
14. $\dfrac{13}{5}$
15. $\dfrac{4}{3}$
16.

17. $\dfrac{2\sqrt{5}}{5}$
18. x
19. $2x^2 - 1$

7.6 Trigonometric Equations

1. $\dfrac{2\pi}{3}, \dfrac{4\pi}{3}$
2. $\dfrac{\pi}{6}, \dfrac{\pi}{2}, \dfrac{5\pi}{6}$

3. $\dfrac{3\pi}{4}, \dfrac{5\pi}{4}$
4. $\dfrac{3\pi}{2}$

5. 30°, 90°, 150°
6. 60°, 300°

7. 90°, 270°
8. 0°, 45°, 180°, 225°

9. 90°, 210°, 330°
10. 0°, 60°, 120°, 180°, 240°, 300°

11. $\dfrac{n \cdot \pi}{2}$
12. $\dfrac{\pi}{4} + \dfrac{n \cdot \pi}{2}, \dfrac{\pi}{2} + n \cdot \pi$

13. $\dfrac{\pi}{6}, \dfrac{\pi}{3}, \dfrac{7\pi}{6}, \dfrac{4\pi}{3}$
14. $0, \dfrac{\pi}{4}, \dfrac{\pi}{2}, \dfrac{3\pi}{4}, \pi, \dfrac{5\pi}{4}, \dfrac{3\pi}{2}, \dfrac{7\pi}{4}$

15. $0, \dfrac{\pi}{6}, \pi, \dfrac{11\pi}{6}$

16. $\dfrac{\pi}{12}, \dfrac{3\pi}{4}, \dfrac{13\pi}{12}, \dfrac{5\pi}{12}, \dfrac{7\pi}{4}, \dfrac{17\pi}{12}$

17. $45°, 225°$

18. $20°, 100°, 140°, 220°, 260°, 340°$

19. $0°, 90°, 270°$

20. $60°, 300°$

7.7 Equations Involving Inverse Trigonometric Functions

1. $x = \dfrac{\sec^{-1}(y-\sqrt{5})}{2}$

2. $x = \sin^{-1}\dfrac{y}{10}$

3. $x = \dfrac{\tan^{-1}(4y)}{7}$

4. $x = \dfrac{\cos^{-1}(y/5)}{4}$

5. $x = 3\sin^{-1}\dfrac{y}{5}$

6. $x = \dfrac{\csc^{-1}(-y/7)}{3}$

7. $x = \sin^{-1} y + 1$

8. $x = \tan^{-1}(y-5)$

9. $x = \cos^{-1}\left(\dfrac{y+5}{3}\right)$

10. $x = \tan^{-1}\left(\dfrac{3-y}{7}\right)$

11. 4

12. $\pi + 6$

13. $\dfrac{5}{13}$

14. 2

15. $\dfrac{\sqrt{10}}{10}$

16. $-\dfrac{\sqrt{3}}{2}$

17. 1

18. -1

19. $\dfrac{1}{2}$

20. $\sqrt{3}$

8.1 The Law of Sines

Solve each triangle *ABC* that exists.

1. $B = 107°$, $C = 11°$, $a = 0.9$

2. $A = 58°$, $a = 42.4$, $b = 50$

3. $A = 54°$, $a = 24$, $c = 17$

4. $A = 29.7°$, $C = 92°$, $b = 16.4$

5. $A = 80°$, $B = 30°$, $b = 14$

6. $A = 110°$, $b = 6$, $c = 7$

7. $C = 12°$, $a = 5$, $c = 5.4$

8. $A = 110°$, $a = 125$, $b = 100$

9. $A = 37°$, $a = 28$, $c = 40$

10. $A = 54°$, $C = 54°$, $b = 25$

11. $A = 33°$, $C = 82°$, $b = 18$

12. $C = 129°$, $a = 10$, $c = 8$

13. $A = 19°$, $C = 102°$, $c = 37$

14. $A = 5.25°$, $b = 4.5$, $c = 22$

15. $B = 25°$, $C = 125°$, $b = 5$

16. $A = 55°$, $a = 19$, $c = 18$

17. $A = 43°$, $B = 71°$, $a = 215$

18. $B = 28.6°$, $a = 41$, $b = 53$

19. $A = 21.3°$, $B = 19.2°$, $a = 143$

20. $A = 60°$, $a = 10$, $b = 12$

8.2 The Law of Cosines

Solve each triangle.

1. $B = 43°$, $a = 91$, $c = 88$

2. $A = 120°$, $b = 3$, $c = 10$

3. $a = 10$, $b = 12$, $c = 14$

4. $A = 56°$, $b = 9$, $c = 3$

5. $a = 48$, $b = 75$, $c = 63$

6. $B = 10.5°$, $a = 6.8$, $c = 2.4$

7. $a = 30$, $b = 40$, $c = 55$

8. $A = 116°$, $B = 34°$, $c = 8.5$

9. $A = 26°$, $B = 64°$, $b = 8$

10. $C = 121°$, $a = 3$, $b = 3$

11. $a = 14.3$, $b = 30$, $c = 21$

12. $C = 15°$, $a = 6.25$, $b = 2.15$

13. $a = 80$, $b = 92$, $c = 124$

14. $A = 22°$, $B = 170°$, $a = 9$

15. $a = 4.4$, $b = 3.8$, $c = 5.2$

Find the area of each triangle ABC.

16. $a = 10$, $b = 12$, $c = 14$

17. $a = 48$, $b = 75$, $c = 63$

18. $a = 30$, $b = 40$, $c = 55$

19. $a = 4.4$, $b = 3.8$, $c = 5.2$

20. $a = 14.3$, $b = 30$, $c = 21$

8.3 Vectors, Operations, and the Dot Product

Given vectors a and b, find $-4\mathbf{a}$ and $\mathbf{a} + 3\mathbf{b}$.

1. $\mathbf{a} = 4i - 3j,\ \mathbf{b} = i - 2j$
2. $\mathbf{a} = \langle -2, 3 \rangle,\ \mathbf{b} = \langle -2, 1 \rangle$

Find the magnitude and direction angle for u.

3. $\langle 6, -6 \rangle$
4. $\langle 2, -1 \rangle$

For each of the following, vector v has the given magnitude and direction. Find the magnitudes of the horizontal and vertical components of v, if α is the angle of inclination of v from the horizontal.

5. $\alpha = 127°,\ |\mathbf{v}| = 5$
6. $\alpha = 297°,\ |\mathbf{v}| = 45$
7. $\alpha = 27°,\ |\mathbf{v}| = 5$
8. $\alpha = 45°,\ |\mathbf{v}| = 4$

Two forces act at a point in the plane. The angle between the two forces is given. Find the magnitude of the resultant force.

9. forces of 2.4 and 1.3 newtons, forming an angle of $120°$

10. forces of 3 and 8 pounds, forming an angle of $80°$

Given $\mathbf{u} = \langle -2, 4 \rangle$ and $\mathbf{v} = \langle -2, 3 \rangle$, find the following.

11. $6\mathbf{u} + 2\mathbf{v}$
12. $2\mathbf{u} + 3\mathbf{v}$

Find the dot product for each pair of vectors.

13. $\langle -1, 2 \rangle,\ \langle 0, 5 \rangle$
14. $3i - 3j,\ 2i + j$

Find the angle between each pair of vectors.

15. $\langle \sqrt{3}, 1 \rangle,\ \langle 3, 3 \rangle$
16. $7i + 8j,\ -2i + 10j$

8.4 Applications of Vectors

Solve each problem.

1. An airplane is traveling with an airspeed of 420 mph and a heading of 72°. A wind of 32 mph is blowing at a heading of 130°. Find the ground speed and the course of the airplane.

2. A 240-pound box is on a 36° ramp. Find the magnitude of the component of the force that is parallel to the ramp.

3. Find the horizontal and vertical components of a 3-pound force that makes an angle of 60° with the x-axis.

4. A boat with a ground speed of 16 mph in still water is crossing a river perpendicular to a current of 6 mph. By what angle is the boat pushed off its course?

5. An airplane is traveling with an airspeed of 270 mph and a heading of 75°. A wind of 72 mph is blowing at a heading of 165°. Find the ground speed of the plane.

6. A 4200-pound car is on a 25° ramp. Find the force required to keep the car from rolling down the ramp.

7. A force of 750 pounds is needed to pull a sailboat and its trailer up a ramp inclined 20° from the horizontal. Find the weight of the sailboat and trailer.

8. Find the magnitude of the resultant force when forces of 170 pounds and 150 pounds act at an angle of 50° to each other.

9. Refer to exercise 8 above. Find the angle between the resultant force and the 170-pound force.

10. Find the amount of force required to pull a 250-pound crate up a ramp with an incline of 30°.

11. An airplane is traveling with an airspeed of 400 mph and a heading of 320°. A wind of 70 mph is blowing at a heading of 190°. Find the ground speed of the airplane.

12. A boat with a groundspeed of 13 mph in still water is crossing a river perpendicular to a current of 7 mph. By what angle is the boat pushed off its course?

13. Find the magnitude of the resultant force when forces of 5 pounds and 7 pounds act at an angle of 50° to each other.

14. A boat with a ground speed of 15 mph in still water is crossing a river perpendicular to a current of 3 mph. By what angle is the boat pushed off its course?

8.5 Trigonometric (Polar) Form of Complex Numbers; Products and Quotients

Find the sum of each pair of complex numbers.

1. $5 - 3i$, $-2i$
2. $-2 + 7i$, $-8 - 3i$
3. 8, $-3i$

Write each complex number in rectangular form.

4. $6(\cos 210° + i \sin 210°)$
5. $5(\cos 315° + i \sin 315°)$
6. $2 cis 120°$
7. $4 cis 210°$

Write each complex number in trigonometric form $r(\cos \theta + i \sin \theta)$ **with** θ **in the interval [0°, 360°).**

8. $\sqrt{3} + i$
9. $6i$
10. 7
11. $-1 + \sqrt{3}i$

Find each product and write it in rectangular form.

12. $[2(\cos 150° + i \sin 150°)][3(\cos 0° + i \sin 0°)]$
13. $(8 cis 240°)(2 cis 300°)$
14. $(4 cis 0°)(2 cis 60°)$

Find each quotient and write it in rectangular form.

15. $\dfrac{2 cis 120°}{4 cis 210°}$
16. $\dfrac{cis 300°}{cis 180°}$
17. $\dfrac{12 cis 60°}{3 cis 120°}$
18. $\dfrac{9 cis 30°}{5 cis 60°}$

Use a calculator to perform the indicated operations. Give answers in rectangular form.

19. $(3 cis 18.75°)(6 cis 230.5°)$
20. $\dfrac{3 cis 18.75°}{6 cis 230.5°}$

8.6 DeMoivre's Theorem; Powers and Roots of Complex Numbers

Find each power. Write each answer in rectangular form.

1. $[2(\cos 15° + i \sin 15°)]^4$

2. $[2(\cos 30° + i \sin 30°)]^5$

3. $[5(\cos 20° + i \sin 20°)]^3$

4. $[3 cis 40°]^3$

5. $(-1 + \sqrt{3}i)^{12}$

6. $2(\sqrt{3} + i)^7$

7. $[cis 0°]^{20}$

Find all cube roots of each complex number. Leave answers in trigonometric form.

8. $8(\cos 135° + i \sin 135°)$

9. $-27i$

10. -1

11. 27

12. $-2 + 2i$

13. $-\dfrac{125}{2}(1 + \sqrt{3}i)$

14. $-4\sqrt{2}(1 - i)$

Find all complex number solutions of each equation. Leave answers in rectangular form.

15. $x^3 - i = 0$

16. $x^3 + 8 = 0$

17. $x^3 - 27 = 0$

18. $x^4 + 1 = 0$

19. $x^3 - 64 = 0$

20. $x^4 - 81 = 0$

21. $x^4 - 16 = 0$

8.7 Polar Equations and Graphs

Convert the following points from polar coordinates to rectangular coordinates.

1. $(2, \frac{4\pi}{3})$

2. $(-\sqrt{2}, 135°)$

3. $(6, 210°)$

4. $(2, \pi)$

Convert the following points from rectangular coordinates to polar coordinates where $0° \leq \theta < 360°$.

5. $(4, -5)$

6. $(-1, \sqrt{3})$

7. $(-3, 6)$

8. $(-1, 1)$

For each rectangular equation, give its equivalent polar form.

9. $x^2 + y^2 - 2x = 0$

10. $x + y = 4$

11. $x^2 + y^2 = 9$

Identify the type of polar graph.

12. $r = 2\sin\theta$

13. $r = 4\sin 2\theta$

14. $r = 2\cos\theta$

15. $r = \theta$

16. $r = 3 + 2\cos\theta$

For each equation, find an equivalent equation in rectangular coordinates.

17. $r = \sec\theta$

18. $r^2 = 9\sin 2\theta$

19. $r = 2\cos\theta$

20. $\theta = \frac{\pi}{3}$

For each plane curve find a rectangular equation for the curve.

1. $x = \sin t$, $y = \cos t$ for t in $[0, 2\pi)$

2. $x = t^2 - 1$, $y = t + 3$ for t in $(-\infty, \infty)$

3. $x = \dfrac{1}{\sqrt{t+1}}$, $y = \dfrac{t}{t+1}$ for t in $(-1, \infty)$

4. $x = 3t - 2$, $y = t + 1$ for t in $(-\infty, \infty)$

5. $x = 2 + 3\cos t$, $y = 3 + 2\sin t$ for t in $[0, 2\pi)$

6. $x = t^2$, $y = -2t^2 + 1$ for t in $(-\infty, \infty)$

7. $x = \dfrac{1}{4}t$, $y = t^2$ for t in $(-\infty, \infty)$

8. $x = 7\sin t$, $y = 7\cos t$ for t in $(-\infty, \infty)$

9. $x = 3 + \sin t$, $y = \cos t - 2$ for t in $(-\infty, \infty)$

10. $x = t^2$, $y = t^3$ for t in $(-\infty, \infty)$

11. $x = \dfrac{1}{2}\sin^2 t$, $y = \cos^2 t$ for t in $(-\infty, \infty)$

12. $x = t - 1$, $y = \dfrac{t}{t-1}$ for t in $(-\infty, \infty)$

13. $x = \tan t$, $y = 2\tan t + 3$ for t in $(-\infty, \infty)$

14. $x = 3 + 2\sec t$, $y = 2 + 4\tan t$ for t in $(-\infty, \infty)$

15. $x = 2 - t^2$, $y = 3 + 2t^2$ for t in $(-\infty, \infty)$

16. $x = 3\sin t$, $y = 4\cos t$ for t in $[0, 2\pi)$

17. $x = t^3$, $y = 3\ln t$ for t in $(0, \infty)$

18. $x = \sqrt{t}$, $y = \sqrt{1-t}$ for t in $(0, 1)$

19. $x = 3t$, $y = 6t + 4$ for t in $[0, 2]$

Answers

8.1 The Law of Sines

1. $A = 62°, b = 0.97, c = 0.19$
2. No Solution
3. $B = 91°, C = 35°, b = 29.66$
4. $B = 58.3°, a = 9.55, c = 19.26$
5. $C = 70°, a = 27.57, c = 26.31$
6. $B = 31.9°, C = 38.1°, a = 10.7$
7. $A = 11.1°, B = 156.9°, b = 10.19$
8. $B = 48.74°, C = 21.26°, c = 48.23$
9. $B = 22.3°, C = 120.7°, b = 17.6$ or $B = 83.7°, C = 59.3°, b = 46.2$
10. $B = 72°, a = 21.3, c = 21.3$
11. $B = 65°, a = 10.8, c = 19.7$
12. No Solution
13. $B = 59°, a = 12.3, b = 32.4$
14. $B = 1.35°, C = 6.6°, a = 17.5$
15. $A = 30°, a = 5.9, c = 9.7$
16. $B = 74.1°, C = 50.9°, b = 22.3$
17. $C = 66°, b = 298, c = 288$
18. $A = 21.7°, C = 129.7°, c = 85.2$
19. $C = 139.5°, b = 129.5, c = 255.7$
20. No Solution

8.2 The Law of Cosines

1. $A = 70.9°, C = 66.1°, b = 65.7$
2. $B = 12.7°, C = 47.3°, a = 11.79$
3. $A = 44.4°, B = 57.1°, C = 78.5°$
4. $B = 105.2°, C = 18.8°, a = 7.7$
5. $A = 39.5°, B = 83.8°, C = 56.6°$
6. $A = 163.9°, C = 5.6°, b = 4.5$
7. $A = 32.2°, B = 45.2°, C = 102.6°$
8. $C = 30°, a = 15.3, b = 9.5$
9. $C = 90°, a = 3.9, c = 8.9$
10. $A = 29.5°, B = 29.5°, c = 5.2$
11. $A = 25.6°, B = 115.1°, C = 39.3°$
12. $A = 157.4°, B = 7.6°, c = 4.21$
13. $A = 40.1°, B = 47.9°, C = 92°$
14. No Solution

15. $A = 56°, B = 45.7°, C = 78.3°$ **16.** 58.79

17. 1503.30 **18.** 585.47

19. 8.19 **20.** 136

8.3 Vectors, Operations, and the Dot Product

1. $-16i + 12j;\ 7i - 9j$ **2.** $\langle 8, -12 \rangle;\ \langle -8, 6 \rangle$ **3.** $6\sqrt{2},\ 315°$ **4.** $\sqrt{5},\ 333.4°$

5. $-3, 4$ **6.** $20.4, 40.1$ **7.** $4.46, 2.27$ **8.** $2\sqrt{2}, 2\sqrt{2}$

9. 2.1 newtons **10.** 9 pounds **11.** $\langle -16, 30 \rangle$ **12.** $\langle -10, 17 \rangle$

13. 10 **14.** 3 **15.** $15°$ **16.** $52.5°$

8.4 Applications of Vectors

1. 438 mph, $75.6°$ **2.** 141 pounds **3.** $\dfrac{3}{2}, \dfrac{3\sqrt{3}}{2}$ **4.** $21°$

5. 279 mph **6.** 1775 pounds **7.** 2193 pounds **8.** 290 pounds

9. $23.3°$ **10.** 125 pounds **11.** 359 mph **12.** $28°$

13. 11 pounds **14.** $11°$

8.5 Trigonometric Form of Complex Numbers; Products and Quotients

1. $5 - 5i$ **2.** $-10 + 4i$ **3.** $8 - 3i$ **4.** $-3\sqrt{3} - 3i$

5. $\dfrac{5\sqrt{2}}{2} - \dfrac{5\sqrt{2}}{2}i$ **6.** $-1 + \sqrt{3}i$ **7.** $-2\sqrt{3} - 2i$ **8.** $2cis\,30°$

9. $6cis\,90°$ **10.** $7cis\,0°$ **11.** $2cis\,120°$ **12.** $-3\sqrt{3} + 3i$

Answers

13. -16
14. $4+4\sqrt{3}i$
15. $-\dfrac{1}{2}i$
16. $-\dfrac{1}{2}+\dfrac{\sqrt{3}}{2}i$

17. $2-2\sqrt{3}i$
18. $\dfrac{9\sqrt{3}}{10}-\dfrac{9}{10}i$
19. $-6.38-16.83i$
20. $-0.43+0.26i$

8.6 De Moivre's Theorem; Powers and Roots of Complex Numbers

1. $8+8\sqrt{3}i$
2. $-16\sqrt{3}+16i$
3. $\dfrac{125}{2}+\dfrac{125\sqrt{3}}{2}i$
4. $-\dfrac{27}{2}+\dfrac{27\sqrt{3}}{2}i$
5. 4096
6. $-128\sqrt{3}-128i$
7. 1
8. $2cis45°$, $2cis165°$, $2cis285°$
9. $3cis90°$, $3cis210°$, $3cis330°$
10. $cis\dfrac{\pi}{3}$, $cis\pi$, $cis\dfrac{5\pi}{3}$
11. $3cis0°$, $3cis120°$, $3cis240°$
12. $\sqrt{2}cis45°$, $\sqrt{2}cis165°$, $\sqrt{2}cis285°$
13. $5cis80°$, $5cis200°$, $5cis320°$
14. $2cis45°$, $2cis165°$, $2cis285°$
15. $\dfrac{\sqrt{3}}{2}+\dfrac{1}{2}i$, $-\dfrac{\sqrt{3}}{2}+\dfrac{1}{2}i$, $-i$
16. $1-\sqrt{3}i$, $1+\sqrt{3}i$, -2
17. 3, $-\dfrac{3}{2}+\dfrac{3\sqrt{3}}{2}i$, $-\dfrac{3}{2}-\dfrac{3\sqrt{3}}{2}i$
18. $\dfrac{\sqrt{2}}{2}+\dfrac{\sqrt{2}}{2}i$, $\dfrac{\sqrt{2}}{2}-\dfrac{\sqrt{2}}{2}i$, $-\dfrac{\sqrt{2}}{2}+\dfrac{\sqrt{2}}{2}i$, $-\dfrac{\sqrt{2}}{2}-\dfrac{\sqrt{2}}{2}i$,
19. $-2+2\sqrt{3}i$, $-2-2\sqrt{3}i$, 4
20. 3, $3i$, -3, $-3i$
21. 2, -2, $2i$, $-2i$

8.7 Polar Equations and Graphs

1. $(-1,-\sqrt{3})$
2. $(1,-1)$
3. $(-3\sqrt{3},-3)$
4. $(-2,0)$
5. $(\sqrt{41},308.7°)$
6. $(2,120°)$
7. $(3\sqrt{5},116.6°)$
8. $(\sqrt{2},135°)$
9. $r=2\cos\theta$
10. $r=\dfrac{4}{\cos\theta+\sin\theta}$
11. $r=3$
12. circle
13. rose
14. circle
15. spiral
16. limacon
17. $x=1$
18. $(x^2+y^2)^2=18xy$
19. $x^2+y^2=2x$
20. $y=\sqrt{3}x$

8.8 Parametric Equations, Graphs, and Applications

1. $x^2+y^2=1$
2. $x+1=(y-3)^2$
3. $y=1-x^2,\ x\geq 0$
4. $3y-x=5$
5. $\dfrac{(x-2)^2}{9}+\dfrac{(y-3)^2}{4}=1$
6. $y=-2x+1, x\geq 0$
7. $y=16x^2$
8. $x^2+y^2=49$
9. $(x-3)^2+(y+2)^2=1$
10. $y=\pm x^{3/2}$

Answers

11. $y = -2x+1$, $0 \leq x \leq \frac{1}{2}$

12. $y = \dfrac{x+1}{x}$

13. $y = 2x+3$

14. $\dfrac{(x-3)^2}{4} - \dfrac{(y-2)^2}{16} = 1$

15. $y = -2x+7$, $x \leq 2$

16. $\dfrac{x^2}{9} + \dfrac{y^2}{16} = 1$

17. $y = \ln x$

18. $y = \sqrt{1-x^2}$, $x \geq 0$

19. $y = 2x+4$, $0 \leq x \leq 6$

9.1 Systems of Linear Equations

Solve each system by substitution.

1. $x + y = 16$
   $y = 7x$

2. $y = 2x - 8$
   $4y + 12 = 2x$

3. $x + y = 4$
   $2x + 3y = 9$

4. $3x + y = 9$
   $2x - 2y = 12$

Solve each system by elimination.

5. $2x + 4y = -12$
   $3x + 2y = 1$

6. $3x + y = 7$
   $2x - 5y = -1$

7. $x - y = 6$
   $2x - 3y = 9$

8. $2x + 3y = 11$
   $3x - y = 4$

Solve each system. State whether it is inconsistent or has infinitely many solutions.

9. $2x - y = 1$
   $8x - 4y = 3$

10. $2x - 5y = 6$
    $4x = 12 + 10y$

11. $6x + 2y = 1$
    $12x = 2 - 4y$

12. $2x + 3y = 2$
    $4x + 6y = 7$

Use a graphing calculator to solve each system. Express solutions to the nearest thousandth.

13. $5.8x - 2.5y = 6.44$
    $3.7x + 2.3y = 10.29$

14. $\sqrt{2}x - 4y = 9$
    $10x + 11y = 12$

15. $0.3x - \sqrt{2}y = 2$
    $\sqrt{7}x + 0.5y = 1$

16. $0.9x + 0.3y = 1.03$
    $2.5x - 8.5y = 11.1$

Solve each system of equations in three variables.

17. $x + 2y + 3z = 9$
    $2x - y + 2z = 11$
    $3x + 4y - 2z = -4$

18. $2x - 2y + 3z = 1$
    $2x - 6y - 4z = -9$
    $x + y + z = 6$

9.2 Matrix Solution of Linear Systems

Write the augmented matrix for each system and give its size. Do not solve the system.

1. $5x - 2y = -10$
   $4x + 3y = -1$

2. $3x + 6y - 3z = -1$
   $-4x - 3y - 4z = 3$
   $2x + 2y + 4z = 2$

Write the system of equations associated with each augmented matrix. Do not solve the system.

3. $\begin{bmatrix} 1 & 1 & -1 & | & 2 \\ 2 & -1 & 1 & | & 6 \\ -1 & 2 & -1 & | & 1 \end{bmatrix}$

4. $\begin{bmatrix} 3 & -4 & 8 & | & 0 \\ 5 & 3 & -4 & | & -2 \\ -5 & 3 & -8 & | & -3 \end{bmatrix}$

Use the Gauss-Jordan method to solve each system of equations.

5. $-x + 4y = 9$
   $3x - 2y = 9$

6. $4x - 6y = 2$
   $2x - 2y = -1$

7. $x + 2y = 9$
   $3x + 4y = -4$

8. $2x - 3y = 6$
   $5x + y = 1$

9. $x + 5y + 7z = 2$
   $3x + 9y - 2z = 3$
   $6y - 4z = 1$

10. $2x + 3y - z = -8$
    $x + 2y + 2z = 3$
    $-3x + y + z = -5$

11. $3x - y - z = -2$
    $2x + 2y + 3z = 3$
    $-3x + 2y + 2z = 5$

12. $-x + y - 4z = 12$
    $-8x - 20y + 4z = 15$
    $2x + 5y - z = 12$

13. $2x - y + 3z = -2$
    $y + 2z = 1$
    $3x - 2y + z = 3$

14. $-x + 2y - 3z = 1$
    $2x + y + z = 0$
    $x + 3y + 2z = 4$

15. $-x - 2y - 3z = 3$
    $2x + 4y + z = 2$
    $-x + 2y = 1$

16. $2x - y + 3z = 2$
    $x + 2y + 3z = -1$
    $3x - 2y + z = 3$

9.3 Determinant Solution of Linear Systems

Find the value of each determinant.

1. $\begin{vmatrix} 1 & 1 \\ -7 & 1 \end{vmatrix}$

2. $\begin{vmatrix} -2 & 1 \\ -2 & 4 \end{vmatrix}$

3. $\begin{vmatrix} 1 & 1 \\ 2 & 3 \end{vmatrix}$

4. $\begin{vmatrix} 3 & 1 \\ 2 & -2 \end{vmatrix}$

5. $\begin{vmatrix} 2 & 4 & -12 \\ 3 & 2 & 1 \\ 3 & 1 & 7 \end{vmatrix}$

6. $\begin{vmatrix} 2 & -5 & -1 \\ 1 & -1 & 6 \\ 2 & -3 & 9 \end{vmatrix}$

7. $\begin{vmatrix} 2 & 3 & 11 \\ 3 & -1 & 4 \\ 2 & -1 & 1 \end{vmatrix}$

8. $\begin{vmatrix} 8 & -4 & 3 \\ 2 & -5 & 6 \\ 6 & 2 & 1 \end{vmatrix}$

Solve each equation for x.

9. $\begin{vmatrix} x & 10 \\ 2 & 3 \end{vmatrix} = 1$

10. $\begin{vmatrix} 3 & 4 \\ 5 & x \end{vmatrix} = 3$

11. $\begin{vmatrix} 1 & 2x & -4 \\ 1 & x & 3 \\ -2 & -1 & -1 \end{vmatrix} = 0$

12. $\begin{vmatrix} x & -2 & 1 \\ 3 & 2 & 2 \\ 3 & -2 & 1 \end{vmatrix} = 0$

Find the area of each triangle having vertices at P, Q, and R.

13. $P(2,-3)$, $Q(-3,-4)$, $R(5,6)$

14. $P(2,3)$, $Q(3,-2)$, $R(-2,4)$

Evaluate each determinant.

15. $\begin{vmatrix} 1 & 2 & -1 & 2 \\ 3 & 0 & 2 & 4 \\ 1 & 0 & -1 & -2 \\ 0 & 3 & 2 & -1 \end{vmatrix}$

16. $\begin{vmatrix} 1 & 0 & 0 & 1 \\ 2 & -1 & 0 & 2 \\ 1 & 2 & 3 & -1 \\ 1 & 2 & -1 & 1 \end{vmatrix}$

Use Cramer's Rule to solve each system of equations.

17. $2x + y = 8$
    $3x + 5y = 10$

18. $3x + 5y = 4$
    $5x + 2y = 9$

19. $x - 3y - 4z = 12$
    $x - 3y + 2z = 0$
    $x - y + 2z = 6$

20. $2x - 5y - 4z = 3$
    $x - 6y + 11z = -6$
    $x - 2y - z = 2$

9.4 Partial Fractions

Find the partial fraction decomposition for each rational expression.

1. $\dfrac{6x+2}{5x(x-3)}$

2. $\dfrac{x-2}{(x-5)^2}$

3. $\dfrac{5x^2+30x+27}{3x(x+3)^2}$

4. $\dfrac{3x^2+x+2}{(x+2)(x+3)(x-1)}$

5. $\dfrac{-x-4}{x(x+1)}$

6. $\dfrac{-x-8}{x(x+4)}$

7. $\dfrac{8x-5}{x(x-1)}$

8. $\dfrac{3-3x}{(x+1)(2x-1)}$

9. $\dfrac{2x^2+20x+26}{(x-1)^2(x+3)^2}$

10. $\dfrac{-2x^2+28x-6}{x(x-3)(x+2)}$

11. $\dfrac{5x-13}{x^2-5x+6}$

12. $\dfrac{9x^2-9x+6}{(2x-1)(x-1)(x+2)}$

13. $\dfrac{2x-5}{x^2-6x+9}$

14. $\dfrac{2x^2+10x+2}{(x+1)(x-2)(x+3)}$

15. $\dfrac{x+12}{(2x-1)(x+2)}$

16. $\dfrac{-7x-14}{3x^2-10x-8}$

17. $\dfrac{18x+16}{3x^2+8x+4}$

18. $\dfrac{6x^2+22x+18}{(x+1)(x+2)(x+3)}$

9.5 Nonlinear Systems of Equations

A non-linear system is given below. Verify that the given points are solutions of the system.

1. $y = x^2 - 4x + 5$; $\left(\dfrac{7}{2}, \dfrac{13}{4}\right)$ and $(2,1)$
   $3x - 2y = 4$

2. $y = 6x^2 - x$; $(0,0)$ and $\left(\dfrac{1}{3}, \dfrac{1}{3}\right)$
   $y = x$

3. $x^2 + y^2 = 4$; $\left(\dfrac{6}{5}, -\dfrac{8}{5}\right)$ and $(0,2)$
   $3x + y = 2$

4. $y = \ln(x+3)$; $(0.828733, 1.34253)$
   $2x + y = 3$

Give all solutions of each nonlinear system of equations.

5. $2x + 3y = 12$
   $4x^2 + 9y^2 = 144$

6. $9x^2 + 16y^2 = 144$
   $9x^2 + 5y^2 = 45$

7. $xy = 4$
   $2x^2 + y^2 = 18$

8. $9x^2 + 16y^2 = 144$
   $9x^2 - 16y^2 = 144$

9. $x^2 + y^2 = 9$
   $4x - 3y = 0$

10. $y = x^2 + 3x + 2$
    $y = 2x + 3$

11. $x^2 + y^2 = 1$
    $y = x - 1$

12. $y = x^2 + 1$
    $2x^2 + 2y = 6$

13. $y = x^2$
    $y = 8 - x^2$

14. $9x^2 - 16y^2 = 144$
    $y^2 - 4x^2 = 1$

15. $xy = 2$
    $x + y = 4$

16. $3x - y = -2$
    $2x^2 - y = 0$

17. $xy = 4$
    $x^2 + y^2 = 8$

18. $x^2 = y + 8$
    $x^2 + y^2 = 64$

19. $y = x + 4$
    $x^2 + y^2 = 16$

20. $x^2 - 2y = 8$
    $x^2 + y^2 = 16$

21. $x^2 + y^2 = 4$
    $-x^2 + y = 8$

9.6 Systems of Inequalities and Linear Programming

Graph each inequality.

1. $2x + 3y > 6$

2. $x + y \leq 5$

3. $2x \geq y + 8$

4. $4y - 3x \leq 12$

5. $x + y \geq 8$

6. $3y \leq 2x + 12$

7. $2x + 2y > 4$

8. $y - 4x \geq -10$

9. $y < 4x^2 - 6$

10. $(x-2)^2 + (y+1)^2 < 4$

Graph the solution set of each system of inequalities.

11. $y < 9 - 3x$
    $x > -3y + 1$

12. $x < 2y - 5$
    $x + 4y \leq -6$

13. $y \geq -x$
    $y \geq -3$

14. $x + y \leq 8$
    $x - y \geq 2$

15. $y < 16 - x^2$
    $y > x^2 - 4$

16. $y \geq 2$
    $x - y > 0$

17. $y \geq 0$
    $x + 3y \geq 6$

18. $y \geq e^x$
    $y \geq 4$

19. $y \leq \left(\dfrac{1}{3}\right)^x$
    $y \geq 3$

20. $y \leq \log x$
    $y \geq x - 2$

9.7 Properties of Matrices

Find the values of the variables for which each statement is true, if possible.

1. $\begin{bmatrix} 3 & x \\ y & 4 \end{bmatrix} = \begin{bmatrix} 3 & -2 \\ -1 & 4 \end{bmatrix}$

2. $\begin{bmatrix} 3 & x \\ y & -3 \end{bmatrix} = \begin{bmatrix} z & -2 \\ 4 & w \end{bmatrix}$

3. $\begin{bmatrix} 1 & 1 & 1 \\ 3 & 2 & -1 \\ 2 & 3 & 4 \end{bmatrix} = \begin{bmatrix} a & 1 & 1 \\ b & c & -1 \\ 2 & 3 & d \end{bmatrix}$

4. $\begin{bmatrix} x-2 & 1 & 1 \\ 2 & 3 & 3 \\ 4 & 5 & 2 \end{bmatrix} = \begin{bmatrix} 4 & -1 & z-1 \\ y+3 & 3 & 3 \\ 4 & 5 & 2 \end{bmatrix}$

5. $\begin{bmatrix} x & y & z \end{bmatrix} = \begin{bmatrix} 6 & 3 \end{bmatrix}$

Perform each operation, if possible.

6. $\begin{bmatrix} 4 & 3 \\ 3 & 2 \end{bmatrix} + \begin{bmatrix} 4 & 1 \\ 2 & 3 \end{bmatrix}$

7. $\begin{bmatrix} 5 & 3 \\ 2 & 3 \end{bmatrix} - \begin{bmatrix} 3 & 0 \\ 1 & 2 \end{bmatrix}$

8. $\begin{bmatrix} 1 & 3 \\ 4 & 2 \end{bmatrix} + \begin{bmatrix} 2 & -1 \\ 3 & 0 \end{bmatrix}$

9. $\begin{bmatrix} 2 & 1 & 0 \\ -1 & 0 & 2 \\ 3 & -1 & 3 \end{bmatrix} + \begin{bmatrix} 2 & 0 & -1 \\ 1 & 2 & 3 \\ -1 & 2 & 1 \end{bmatrix}$

10. $\begin{bmatrix} 2 & 0 & 4 \\ -1 & 1 & -2 \\ 3 & 0 & 2 \end{bmatrix} - \begin{bmatrix} -1 & 2 & 1 \\ 2 & 1 & 3 \\ 5 & 2 & 0 \end{bmatrix}$

11. $\begin{bmatrix} -x & 3y & 2x \\ 1+x & 2x & 0 \\ x-2 & y-1 & -3y \end{bmatrix} - \begin{bmatrix} x & x & x \\ -1 & 2y & -3 \\ x+2 & 2y & 5y \end{bmatrix}$

Let $A = \begin{bmatrix} 2 & 3 \\ 1 & -1 \end{bmatrix}$ and $B = \begin{bmatrix} 4 & 5 \\ 0 & 7 \end{bmatrix}$. Find each of the following.

12. $3A$

13. $\frac{1}{2}B$

14. $A - 2B$

15. $3A + 2B$

Find each matrix product.

16. $\begin{bmatrix} 1 & 2 \\ 2 & -1 \end{bmatrix} \begin{bmatrix} 3 & 4 \\ 1 & -3 \end{bmatrix}$

17. $\begin{bmatrix} 1 & 2 & 3 \\ 0 & 0 & 1 \end{bmatrix} \begin{bmatrix} 0 & 1 \\ -1 & 2 \\ 2 & 1 \end{bmatrix}$

18. $\begin{bmatrix} 5 & -1 & 11 \\ -3 & -2 & -2 \\ 2 & 4 & 1 \end{bmatrix} \begin{bmatrix} 2 & 1 & 3 \\ 3 & -1 & 2 \\ 4 & 1 & 2 \end{bmatrix}$

19. $\begin{bmatrix} 0 & 1 & 0 \\ -1 & 4 & 3 \\ 2 & 1 & 0 \end{bmatrix} \begin{bmatrix} 3 & -1 \\ 2 & 0 \\ -1 & 1 \end{bmatrix}$

9.8 Matrix Inverses

Decide whether or not the given matrices are inverses of each other.

1. $\begin{bmatrix} 2 & -3 \\ 4 & -7 \end{bmatrix}$ and $\begin{bmatrix} 3.5 & -1.5 \\ 2 & -1 \end{bmatrix}$

2. $\begin{bmatrix} 1 & 1 \\ 3 & 3 \end{bmatrix}$ and $\begin{bmatrix} 7 & -2 \\ -6 & 2 \end{bmatrix}$

3. $\begin{bmatrix} 1 & 5 & 2 \\ 1 & 1 & 7 \\ 0 & -3 & 4 \end{bmatrix}$ and $\begin{bmatrix} -25 & 26 & -33 \\ 4 & -4 & 5 \\ 3 & -3 & 4 \end{bmatrix}$

4. $\begin{bmatrix} 3 & 0 & 0 \\ 3 & 2 & 0 \\ 3 & 2 & 5 \end{bmatrix}$ and $\begin{bmatrix} 1/3 & 0 & 0 \\ -1/2 & 1/2 & 0 \\ 0 & -1/5 & 1/5 \end{bmatrix}$

Find the inverse for each matrix.

5. $\begin{bmatrix} 5 & -2 \\ 3 & 7 \end{bmatrix}$

6. $\begin{bmatrix} -7 & 3 \\ 5 & -2 \end{bmatrix}$

7. $\begin{bmatrix} 1 & -1 & 3 \\ 2 & 1 & 2 \\ -2 & -2 & 1 \end{bmatrix}$

8. $\begin{bmatrix} 0 & 5 & 3 \\ 1 & 2 & 1 \\ 0 & 3 & 2 \end{bmatrix}$

Solve each system by using the inverse of the coefficient matrix.

9. $x + y = 4$
   $2x + 3y = 9$

10. $2x + y = 7$
    $x - y = -4$

11. $7x + 8y = 20$
    $3x + 5y = 12$

12. $x + 2y = 3$
    $2x + y = 1$

13. $x + 2y - 3z = -4$
    $2x - y + z = 3$
    $3x + 2y + z = 10$

14. $x - y = 4$
    $x - z = 0$
    $-6x + 2y + 3z = 7$

Use a graphing calculator to find the inverse of each matrix.

15. $\begin{bmatrix} \sqrt{7} & -2 \\ 1/4 & \sqrt{3} \end{bmatrix}$

16. $\begin{bmatrix} 0.1 & -0.9 & 0.3 \\ 0.2 & 0.4 & 0.2 \\ 0.3 & -0.6 & 0.9 \end{bmatrix}$

Use a graphing calculator and the method of matrix inverses to solve each system.

17. $3.1x + \sqrt{2}y = 6$
    $\sqrt{3}x - 6y = 4$

18. $\sqrt{3}x + \pi y + ez = 2$
    $ex + \pi y + \sqrt{3}z = 1$
    $\pi x + \sqrt{3}y + ez = 3$

Answers

9.1 Systems of Linear Equations

1. $(2, 14)$
2. $\left(\dfrac{10}{3}, -\dfrac{4}{3}\right)$
3. $(3, 1)$
4. $\left(\dfrac{15}{4}, -\dfrac{9}{4}\right)$
5. $\left(\dfrac{7}{2}, -\dfrac{19}{4}\right)$
6. $(2, 1)$
7. $(9, 3)$
8. $\left(\dfrac{23}{11}, \dfrac{25}{11}\right)$
9. $\varnothing$, inconsistent
10. $\left(\dfrac{5y+6}{2}, y\right)$, infinitely many solutions
11. $\left(\dfrac{1-2y}{6}, y\right)$, infinitely many solutions
12. $\varnothing$, inconsistent
13. $(1.794, 1.587)$
14. $(2.646, -1.315)$
15. $(0.620, -1.283)$
16. $(1.439, -0.883)$
17. $(2, -1, 3)$
18. $\left(\dfrac{109}{32}, \dfrac{87}{32}, -\dfrac{1}{8}\right)$

9.2 Matrix Solution of Linear Systems

1. $\begin{bmatrix} 5 & -2 & | & -10 \\ 4 & 3 & | & -1 \end{bmatrix}$

2. $\begin{bmatrix} 3 & 6 & -3 & | & -1 \\ -4 & -3 & -4 & | & 3 \\ 2 & 2 & 4 & | & 2 \end{bmatrix}$

3. $\begin{array}{l} x + y - z = 2 \\ 2x - y + z = 6 \\ -x + 2y - 3z = 1 \end{array}$

4. $\begin{array}{l} 3x - 4y + 8z = 0 \\ 5x + 3y - 4z = -2 \\ -5x + 3y - 8z = -3 \end{array}$

5. $\left(\dfrac{27}{5}, \dfrac{18}{5}\right)$

6. $\left(-\dfrac{5}{2}, -2\right)$

7. $\left(-22, \dfrac{31}{2}\right)$

8. $\left(\dfrac{9}{17}, -\dfrac{28}{17}\right)$

9. $\left(\dfrac{65}{162}, \dfrac{35}{162}, \dfrac{2}{27}\right)$

10. $\left(\dfrac{13}{7}, -\dfrac{39}{14}, \dfrac{47}{14}\right)$

11. $\left(\dfrac{1}{3}, \dfrac{20}{3}, -\dfrac{11}{3}\right)$

12. $\varnothing$

13. $\left(6, \dfrac{31}{5}, -\dfrac{13}{5}\right)$

14. $\left(-\dfrac{19}{20}, \dfrac{23}{20}, \dfrac{3}{4}\right)$

15. $\left(\dfrac{2}{5}, \dfrac{7}{10}, -\dfrac{8}{5}\right)$

Answers

16. $\left(\dfrac{3}{8}, -\dfrac{7}{8}, \dfrac{1}{8}\right)$

9.3 Determinant Solution of Linear Systems

1. 8   **2.** –6   **3.** 1   **4.** –8

5. –10   **6.** 4   **7.** 10   **8.** –170

9. 7   **10.** $\dfrac{23}{3}$ **11.** $\dfrac{7}{19}$ **12.** 3

13. 21   **14.** 9.5   **15.** –110   **16.** 2

17. $\left(\dfrac{30}{7}, -\dfrac{4}{7}\right)$ **18.** $\left(\dfrac{37}{19}, -\dfrac{7}{19}\right)$ **19.** $(13, 3, -2)$ **20.** $\left(\dfrac{23}{5}, \dfrac{7}{5}, -\dfrac{1}{5}\right)$

9.4 Partial Fractions

1. $-\dfrac{2}{15x} + \dfrac{4}{3(x-3)}$ **2.** $\dfrac{1}{x-5} + \dfrac{3}{(x-5)^2}$ **3.** $\dfrac{1}{x} + \dfrac{2}{3(x+3)} + \dfrac{2}{(x+3)^2}$

4. $\dfrac{-4}{x+2} + \dfrac{13}{2(x+3)} + \dfrac{1}{2(x-1)}$ **5.** $-\dfrac{4}{x} + \dfrac{3}{x+1}$ **6.** $-\dfrac{2}{x} + \dfrac{1}{x+4}$

7. $\dfrac{5}{x} + \dfrac{3}{x-1}$ **8.** $-\dfrac{2}{x+1} + \dfrac{1}{2x-1}$ **9.** $\dfrac{3}{(x-1)^2} - \dfrac{1}{(x+3)^2}$

10. $\dfrac{1}{x} + \dfrac{4}{(x-3)} - \dfrac{7}{(x+2)}$ **11.** $\dfrac{3}{x-2} + \dfrac{2}{x-3}$ **12.** $\dfrac{2}{x-1} - \dfrac{3}{2x-1} + \dfrac{4}{x+2}$

13. $\dfrac{2}{x-3} + \dfrac{1}{(x-3)^2}$ **14.** $\dfrac{1}{x+1} + \dfrac{2}{x-2} - \dfrac{1}{x+3}$ **15.** $\dfrac{5}{2x-1} - \dfrac{2}{x+2}$

16. $\dfrac{2}{3x+2} - \dfrac{3}{x-4}$ **17.** $\dfrac{3}{3x+2} + \dfrac{5}{x+2}$ **18.** $\dfrac{1}{x+1} + \dfrac{2}{x+2} + \dfrac{3}{x+3}$

9.5 Nonlinear Systems of Equations

1. yes 2. yes 3. yes

4. yes 5. $(6,0); (0,4)$ 6. $(0,3), (0,-3)$

7. $(2\sqrt{2},\sqrt{2}); (-2\sqrt{2},-\sqrt{2}); (\sqrt{2},2\sqrt{2}); (-\sqrt{2},-2\sqrt{2})$ 8. $(4,0); (-4,0)$

9. $\left(\dfrac{9}{5},\dfrac{12}{5}\right); \left(-\dfrac{9}{5},-\dfrac{12}{5}\right)$ 10. $\left(\dfrac{-1-\sqrt{5}}{2}, 2-\sqrt{5}\right); \left(\dfrac{-1+\sqrt{5}}{2}, 2+\sqrt{5}\right)$

11. $(1,0); (0,-1)$ 12. $(1,2); (-1,2)$

13. $(2,4); (-2,4)$ 14. $\varnothing$

15. $(2-\sqrt{2},2+\sqrt{2}); (2+\sqrt{2},2-\sqrt{2})$ 16. $\left(-\dfrac{1}{2},\dfrac{1}{2}\right); (2,8)$

17. $(2,2); (-2,-2)$ 18. $(-\sqrt{15},7); (\sqrt{15},7); (0,-8)$

19. $(0,4); (-4,0)$ 20. $(0,-4); (2\sqrt{3},2); (-2\sqrt{3},2)$

21. $\varnothing$

9.6 Systems of Inequalities and Linear Programming

1.
2.
3.
4.
5.
6.

108   Answers

7.
8.
9.
10.
11.
12.
13.
14.
15.
16.
17.
18.h
19.
20.

9.7 Properties of Matrices

1. $x = -2, y = -1$

2. $x = -2, y = 4, w = -3, z = 3$

3. $a = 1, b = 3, c = 2, d = 4$

4. $x = 6, y = -1, z = 2$

5. not possible

6. $\begin{bmatrix} 8 & 4 \\ 5 & 5 \end{bmatrix}$

7. $\begin{bmatrix} 2 & 3 \\ 1 & 1 \end{bmatrix}$

8. $\begin{bmatrix} 3 & 2 \\ 7 & 2 \end{bmatrix}$

9. $\begin{bmatrix} 4 & 1 & -1 \\ 0 & 2 & 5 \\ 2 & 1 & 4 \end{bmatrix}$

Answers

10. $\begin{bmatrix} 3 & -2 & 3 \\ -3 & 0 & -5 \\ -2 & -2 & 2 \end{bmatrix}$

11. $\begin{bmatrix} -2x & 3y-x & x \\ 2+x & 2x-2y & 3 \\ -4 & -y-1 & -8y \end{bmatrix}$

12. $\begin{bmatrix} 6 & 9 \\ 3 & -3 \end{bmatrix}$

13. $\begin{bmatrix} 2 & 5/2 \\ 0 & 7/2 \end{bmatrix}$

14. $\begin{bmatrix} -6 & -7 \\ 1 & -15 \end{bmatrix}$

15. $\begin{bmatrix} 14 & 19 \\ 3 & 11 \end{bmatrix}$

16. $\begin{bmatrix} 5 & -2 \\ 5 & 11 \end{bmatrix}$

17. $\begin{bmatrix} 4 & 8 \\ 2 & 1 \end{bmatrix}$

18. $\begin{bmatrix} 51 & 17 & 35 \\ -20 & -3 & -17 \\ 20 & -1 & 16 \end{bmatrix}$

19. $\begin{bmatrix} 2 & 0 \\ 2 & 4 \\ 8 & -2 \end{bmatrix}$

9.8 Matrix Inverses

1. yes

2. no

3. yes

4. yes

5. $\begin{bmatrix} 7/41 & 2/41 \\ -3/41 & 5/41 \end{bmatrix}$

6. $\begin{bmatrix} 2 & 3 \\ 5 & 7 \end{bmatrix}$

7. $\begin{bmatrix} 1 & -1 & -1 \\ -6/5 & 7/5 & 4/5 \\ -2/5 & 4/5 & 3/5 \end{bmatrix}$

8. $\begin{bmatrix} -1 & 1 & 1 \\ 2 & 0 & -3 \\ -3 & 0 & 5 \end{bmatrix}$

9. $(3,1)$

10. $(1,5)$

11. $\left(\dfrac{4}{11}, \dfrac{24}{11}\right)$

12. $\left(-\dfrac{1}{3}, \dfrac{5}{3}\right)$

13. $(1,2,3)$

14. $(-15,-19,-15)$

15. $\begin{bmatrix} 0.341 & 0.394 \\ -0.049 & 0.521 \end{bmatrix}$

16. $\begin{bmatrix} 5.714 & 7.5 & -3.571 \\ -1.429 & 0 & 0.476 \\ -2.857 & -2.5 & 2.619 \end{bmatrix}$

17. $(1.979, -0.095)$

18. $(0.194, -0.515, 1.208)$

10.1 Parabolas

Graph each horizontal parabola.

1. $y^2 = 4x$

2. $2y^2 = -10x$

3. $2x = y^2 - 8y + 12$

4. $x = y^2 + y - 2$

5. $x + 5 = -(y-1)^2$

6. $3(x-2) = (y+4)^2$

Give the focus, directrix, and axis for each parabola.

7. $y^2 + 10x = 0$

8. $8x = y^2 - 2y - 15$

9. $x^2 - 4x = -2y - 6$

10. $y = 3(x+1)^2 - 2$

11. $x^2 + 6x + 8y + 25 = 0$

12. $-10(y-3) = (x+2)^2$

Write an equation for each parabola with vertex at the origin.

13. focus: $(4,0)$

14. focus: $(2,0)$

15. focus: $\left(0, -\frac{3}{2}\right)$

16. focus: $(0,2)$

17. through $(2,4)$, opening up

18. through $(2,4)$, opening to the right

Write an equation for each parabola.

19. vertex: $(-2,3)$, focus: $(-2,5)$

20. vertex: $(5,4)$, focus: $(2,4)$

Determine the two equations necessary to graph each horizontal parabola using a graphing calculator.

21. $(y-2)^2 = 8(x+1)$

22. $x = \frac{1}{4}(y-2)^2 - 1$

10.2 Ellipses

Identify the domain, range, center, vertices, endpoints of the minor axis, and the foci in each figure.

1. $\dfrac{x^2}{9} + \dfrac{y^2}{4} = 1$

2. $9x^2 + 4y^2 = 36$

3. $16x^2 + 25y^2 = 400$

4. $\dfrac{(x+2)^2}{9} + \dfrac{(y+2)^2}{4} = 1$

5. $(x-3)^2 + \dfrac{(y-5)^2}{9} = 1$

6. $9x^2 + 16y^2 = 144$

7. $\dfrac{(x-1)^2}{9} + \dfrac{(y+1)^2}{4} = 1$

8. $y^2 + 36x^2 = 36$

Write an equation for each ellipse.

9. center: $(0,0)$; focus: $(3,0)$; vertex: $(5,0)$

10. foci: $(0,2),(0,-2)$; vertices: $(0,5),(0,-5)$

11. x-intercepts: $(4,0),(-4,0)$; y-intercepts: $(0,3),(0,-3)$

12. x-intercepts: $(3,0),(-3,0)$; y-intercepts: $(0,4),(0,-4)$

13. foci: $(3,0),(-3,0)$; vertices: $(-7,0),(7,0)$

14. vertices: $(0,-2),(0,2)$; length of minor axis $= 2$

15. vertices: $(0,-8),(0,8)$; length of minor axis $= 10$

16. center: $(0,0)$; length of horizontal major axis $= 8$; length of minor axis $= 4$

Graph each equation. Give the domain and range. Identify any that are graphs of functions.

17. $x = \sqrt{1 - \dfrac{y^2}{16}}$

18. $y = -\sqrt{1 - \dfrac{x^2}{9}}$

Determine the two equations necessary to graph each ellipse with a graphing calculator.

19. $\dfrac{x^2}{4} + y^2 = 1$

20. $\dfrac{x^2}{6} + \dfrac{y^2}{10} = 1$

Find the eccentricity of each ellipse.

21. $\dfrac{x^2}{5} + \dfrac{y^2}{9} = 1$

22. $3x^2 + 4y^2 = 12$

10.3 Hyperbolas

Give the domain, range, center, vertices, foci, and equations of the asymptotes.

1. $\dfrac{x^2}{25} - \dfrac{y^2}{9} = 1$

2. $\dfrac{(x-1)^2}{4} - \dfrac{(y-3)^2}{4} = 1$

3. $x^2 - y^2 = 1$

4. $25y^2 - 4x^2 = 100$

5. $\dfrac{(y-1)^2}{4} - \dfrac{(x+3)^2}{16} = 1$

6. $\dfrac{y^2}{9} - x^2 = 1$

7. $2x^2 - y^2 = 4$

Graph each equation. Give the domain and range. Identify any that are the graphs of functions.

8. $y = \sqrt{\dfrac{x^2}{9} - 1}$

9. $x = -\sqrt{1 + \dfrac{y^2}{9}}$

Find the eccentricity of each hyperbola.

10. $y^2 - \dfrac{x^2}{4} = 1$

11. $\dfrac{x^2}{8} - \dfrac{y^2}{4} = 1$

Write an equation for each hyperbola.

12. center: $(1,4)$; focus: $(-2,4)$; vertex: $(0,4)$

13. vertices: $(2,-1),(10,-1)$; foci: $(0,-1),(12,-1)$

14. vertices: $(3,0),(-3,0)$; foci: $(5,0),(-5,0)$

15. y-intercepts: $(0,4),(0,-4)$; foci: $(0,5),(0,-5)$

16. vertices: $(0,3),(0,-3)$; foci: $(0,5),(0,-5)$

17. vertices: $(2,0),(6,0)$; foci: $(0,0),(8,0)$

18. center: $(-3,1)$; focus: $(-3,6)$; vertex: $(-3,4)$

Determine the two equations necessary to graph each hyperbola with a graphing calculator.

19. $3y^2 - 2x^2 = 12$

20. $3x^2 - y^2 = 6x$

10.4 Summary of the Conic Sections

The equation of a conic section is given in a familiar form. Identify the type of graph that each equation has.

1. $\dfrac{y^2}{25} + \dfrac{x^2}{16} = 1$

2. $y^2 - x^2 + 2x = 2$

3. $\dfrac{x^2}{4} - \dfrac{y^2}{16} = 1$

4. $y^2 = 5x$

5. $y^2 + x^2 = 9$

6. $4y^2 - 9x^2 = 36$

7. $2x^2 + 6y^2 = 12$

8. $\dfrac{x^2}{3} + \dfrac{y^2}{2} = 1$

For each equation that has a graph, identify the type of graph.

9. $4x^2 + y^2 - 16x - 6y = 15$

10. $x^2 - 8x + 6y = 0$

11. $y^2 - 8y - 8x^2 + 4x = 8$

12. $x^2 - 8x = 4y$

13. $x^2 - 2x - y^2 - 2y = 1$

14. $x^2 + 16y = 0$

15. $5x^2 + 3y^2 = 15$

16. $x^2 + 2x - 4y^2 = 3$

17. $6y^2 - 24x = 0$

18. $2y^2 + 8y = 9x^2$

19. $3x^2 + 4y^2 = 12$

20. $4x^2 - 6y^2 - 8x + 24y - 44 = 0$

10.1 Parabolas

1.
2.
3.

4.
5.
6.

7. $\left(-\frac{5}{2},0\right); x=\frac{5}{2}; y=0$

8. $(0,1); x=-4; y=1$

9. $\left(2,-\frac{3}{2}\right); y=-\frac{1}{2}; x=2$

10. $\left(-1,-\frac{23}{12}\right); y=-\frac{25}{12}; x=-1$

11. $(-3,-4); y=0; x=-3$

12. $\left(-2,\frac{1}{2}\right); y=\frac{11}{2}; x=-2$

13. $y^2 = 16x$

14. $y^2 = 8x$

15. $x^2 = -6y$

16. $x^2 = 8y$

17. $x^2 = y$

18. $y^2 = 8x$

19. $(x+2)^2 = 8(y-3)$

20. $(y-4)^2 = -12(x-5)$

21. $y = 2 + 2\sqrt{2(x+1)};$
    $y = 2 - 2\sqrt{2(x+1)}$

22. $y = 2 + 2\sqrt{x+1}$
    $y = 2 - 2\sqrt{x+1}$

10.2 Ellipses

1. $[-3,3]; [-2,2]; (0,0); (3,0),(-3,0); (0,2),(0,-2); (\sqrt{5},0),(-\sqrt{5},0)$

2. $[-2,2]; [-3,3]; (0,0); (0,3),(0,-3); (2,0),(-2,0); (0,\sqrt{5}),(0,-\sqrt{5})$

3. $[-5,5]; [-4,4]; (0,0); (5,0),(-5,0); (0,4),(0,-4); (3,0),(-3,0)$

Answers 115

4. $[-5,1]$; $[-4,0]$; $(-2,-2)$; $(-5,-2),(1,-2)$; $(-2,0),(-2,-4)$; $(-2+\sqrt{5},-2),(-2-\sqrt{5},-2)$

5. $[2,4]$; $[2,8]$; $(3,5)$; $(3,8),(3,2)$; $(2,5),(4,5)$; $(3,5+2\sqrt{2}),(3,5-2\sqrt{2})$

6. $[-4,4]$; $[-3,3]$; $(0,0)$; $(4,0),(-4,0)$; $(0,3),(0,-3)$; $(\sqrt{7},0),(-\sqrt{7},0)$

7. $[-2,4]$; $[-3,1]$; $(1,-1)$; $(-2,-1),(4,-1)$; $(1,1),(1,-3)$; $(1+\sqrt{5},-1),(1-\sqrt{5},-1)$

8. $[-1,1]$; $[-6,6]$; $(0,0)$; $(0,6),(0,-6)$; $(1,0),(-1,0)$; $(0,\sqrt{35}),(0,-\sqrt{35})$

9. $\dfrac{x^2}{25}+\dfrac{y^2}{16}=1$ **10.** $\dfrac{x^2}{21}+\dfrac{y^2}{25}=1$ **11.** $\dfrac{x^2}{16}+\dfrac{y^2}{9}=1$

12. $\dfrac{x^2}{9}+\dfrac{y^2}{16}=1$ **13.** $\dfrac{x^2}{49}+\dfrac{y^2}{40}=1$ **14.** $x^2+\dfrac{y^2}{4}=1$

15. $\dfrac{x^2}{25}+\dfrac{y^2}{64}=1$ **16.** $\dfrac{x^2}{16}+\dfrac{y^2}{4}=1$ **17.** $[0,1]$; $[-4,4]$

18. $[-3,3]$; $[-1,0]$; function

19. $y=\sqrt{1-\dfrac{x^2}{4}}$; $y=-\sqrt{1-\dfrac{x^2}{4}}$

20. $y=\sqrt{10\left(1-\dfrac{x^2}{6}\right)}$; $y=-\sqrt{10\left(1-\dfrac{x^2}{6}\right)}$

21. $\dfrac{2}{3}$ **22.** $\dfrac{1}{2}$

10.3 Hyperbolas

1. $(-\infty,-5]\cup[5,\infty)$; $(-\infty,\infty)$; $(0,0)$; $(5,0),(-5,0)$; $(\sqrt{34},0),(-\sqrt{34},0)$; $y=\pm\dfrac{3}{5}x$

2. $(-\infty,-1]\cup[3,\infty)$; $(-\infty,\infty)$; $(1,3)$; $(-1,3),(3,3)$; $(1+2\sqrt{2},3),(1-2\sqrt{2},3)$; $y=\pm x$

3. $(-\infty,-1]\cup[1,\infty)$; $(-\infty,\infty)$; $(0,0)$; $(1,0),(-1,0)$; $(\sqrt{2},0),(-\sqrt{2},0)$; $y=\pm x$

4. $(-\infty,\infty)$; $(-\infty,-2]\cup[2,\infty)$; $(0,0)$; $(0,2),(0,-2)$; $(0,\sqrt{29}),(0,-\sqrt{29})$; $y=\pm\dfrac{2}{5}x$

Answers

5. $(-\infty,\infty); (-\infty,-1]\cup[3,\infty); (-3,1); (-3,3),(-3,-1); (-3,1+2\sqrt{5}),(-3,1-2\sqrt{5}); y-1=\pm\frac{1}{2}(x+3)$

6. $(-\infty,\infty); (-\infty,-3]\cup[3,\infty); (0,0); (0,3),(0,-3); (0,\sqrt{10}),(0,-\sqrt{10}); y=\pm 3x$

7. $(-\infty,-\sqrt{2}]\cup[\sqrt{2},\infty); (-\infty,\infty); (0,0); (\sqrt{2},0),(-\sqrt{2},0); (\sqrt{6},0),(-\sqrt{6},0); y=\pm\sqrt{2}x$

8. $(-\infty,-3]\cup[3,\infty); [0,\infty);$ function

9. $(-\infty,-1]; (-\infty,\infty)$

10. $\sqrt{5}$

11. $\frac{\sqrt{6}}{2}$

12. $(x-1)^2-\frac{(y-4)^2}{8}=1$

13. $\frac{(x-6)^2}{16}-\frac{(y+1)^2}{20}=1$

14. $\frac{x^2}{9}-\frac{y^2}{16}=1$

15. $\frac{y^2}{16}-\frac{x^2}{9}=1$

16. $\frac{y^2}{9}-\frac{x^2}{16}=1$

17. $\frac{(x-4)^2}{4}-\frac{y^2}{12}=1$

18. $\frac{(y-1)^2}{9}-\frac{(x+3)^2}{16}=1$

19. $y=\pm\sqrt{\frac{2}{3}x^2+4}$

20. $y=\pm\sqrt{3x^2-6x}$

10.4 Summary of the Conic Sections

1. ellipse   **2.** hyperbola   **3.** hyperbola   **4.** parabola

5. ellipse   **6.** hyperbola   **7.** ellipse   **8.** ellipse

9. ellipse   **10.** parabola   **11.** hyperbola   **12.** parabola

13. hyperbola   **14.** parabola   **15.** ellipse   **16.** hyperbola

17. parabola   **18.** hyperbola   **19.** ellipse   **20.** hyperbola

11.1 Sequences and Series

Write the first five terms of each sequence.

1. $a_n = n(n+1)$

2. $a_n = \dfrac{n^2 - 3}{3}$

3. $a_n = (-1)^n$

4. $a_n = 2^n - 1$

Find the first four terms of each sequence.

5. $a_1 = 4$
   $a_n = a_{n-1} + 3$ for $n > 1$

6. $a_1 = 5$
   $a_n = a_{n-1} - 3$ for $n > 1$

7. $a_1 = 2$
   $a_n = n \cdot a_{n-1}$ for $n > 1$

8. $a_1 = 81$
   $a_n = \dfrac{1}{3} a_{n-1}$ for $n > 1$

Evaluate each series.

9. $\sum_{i=1}^{5} i^2$

10. $\sum_{i=1}^{8} (7i - 6)$

11. $\sum_{i=1}^{5} \dfrac{i}{i+1}$

12. $\sum_{i=0}^{3} 3 \cdot 4^i$

13. $\sum_{i=1}^{10} (i - 2)$

14. $\sum_{i=1}^{3} \dfrac{1}{i^3}$

Use a graphing calculator to evaluate each series.

15. $\sum_{i=0}^{20} 100(1.06)^i$

16. $\sum_{i=0}^{10} 3 \cdot 2^i$

Write the terms for each series. Evaluate the sum given that $x_1 = -4$, $x_2 = -2$, $x_3 = 0$, $x_4 = 2$, and $x_5 = 4$.

17. $\sum_{i=1}^{5} \dfrac{2}{x_i + 1}$

18. $\sum_{i=1}^{5} \dfrac{3 + 4x_i}{5}$

Use the summation properties and rules to evaluate each series.

19. $\sum_{i=1}^{10} 3$

20. $\sum_{i=1}^{10} 6i$

21. $\sum_{i=1}^{5} 3i^2$

22. $\sum_{i=1}^{10} i^2 + 2i + 2$

11.2 Arithmetic Sequences and Series

Find the common difference d for each arithmetic sequence.

1. 3, 1, −1, −1,...

2. 6, 7.5, 9, 10.5,...

3. 0.25, 0.5, 0.75, 1, ...

4. $x+2, 2x+4, 3x+6, 4x+8, ...$

Write the first five terms of each arithmetic sequence.

5. $a_1 = 7; d = 4$

6. $a_1 = 9; d = \dfrac{1}{3}$

Find a_5 and a_n for each arithmetic sequence.

7. 65, 80, 95, ...

8. −6, −1, 4, ...

9. $a_1 = 2, a_4 = 8$

10. $a_2 = -5, a_4 = 5$

Evaluate S_8, the sum of the first eight terms of each arithmetic sequence.

11. 24, 16, 8, 0, ...

12. $2, \dfrac{5}{2}, 3, \dfrac{7}{2}, ...$

13. $a_1 = 3, a_4 = 9$

14. $a_2 = 1, a_5 = -8$

Find a_1 and d for each arithmetic series.

15. $S_7 = -6, a_2 = 4$

16. $S_3 = -60, a_2 = -20$

Evaluate each sum.

17. $\sum_{i=1}^{5}(i+1)$

18. $\sum_{i=1}^{10}(2i+2)$

19. $\sum_{i=1}^{8}(2i-1)$

20. $\sum_{i=1}^{6}(5i-2)$

11.3 Geometric Sequences and Series

Find a_4 and a_n for each geometric sequence.

1. 8, 4, 2, ...

2. −2, 6, −18, ...

3. $a_1 = 1$, $r = 4$

4. $a_1 = 9$, $r = \dfrac{1}{3}$

Find a_1 and r for each geometric sequence.

5. $a_2 = 9$, $a_5 = \dfrac{1}{3}$

6. $a_3 = -6$, $a_9 = -\dfrac{3}{32}$

Use the formula for S_n to find the sum of the first five terms of each geometric sequence.

7. 20, 5, $\dfrac{5}{4}$, $\dfrac{5}{16}$, ...

8. $\dfrac{2}{5}$, $\dfrac{4}{5}$, $\dfrac{8}{5}$, ...

9. $a_1 = 2$, $r = 5$

10. $a_1 = 81$, $r = -\dfrac{1}{3}$

Find each sum.

11. $\displaystyle\sum_{i=1}^{5}\left(\dfrac{1}{5}\right)^i$

12. $\displaystyle\sum_{i=1}^{3}\left(-\dfrac{3}{10}\right)^i$

13. $\displaystyle\sum_{i=1}^{5} 2\cdot 3^i$

14. $\displaystyle\sum_{i=1}^{3} 5\cdot\left(\dfrac{3}{4}\right)^i$

Find r for each infinite geometric sequence. Identify any whose sum does not converge.

15. 2, −6, 18, ...

16. 5, 1, $\dfrac{1}{5}$, $\dfrac{1}{25}$, ...

Evaluate each sum that converges.

17. $\displaystyle\sum_{i=1}^{\infty}\left(\dfrac{1}{3}\right)^i$

18. $\displaystyle\sum_{i=1}^{\infty} 10(0.1)^{i-1}$

19. $\displaystyle\sum_{i=1}^{\infty} 5(1.07)^i$

20. $\displaystyle\sum_{i=1}^{\infty} 2\left(\dfrac{3}{5}\right)^i$

11.4 The Binomial Theorem

Evaluate each expression.

1. $\dfrac{9!}{5!4!}$

2. $\dfrac{5!}{3!2!}$

3. $\dfrac{6!}{2!4!}$

4. $\binom{6}{3}$

5. $\binom{4}{3}$

6. $\binom{7}{2}$

7. $_5C_5$

8. $_7C_3$

9. $_6C_4$

Write the binomial expansion for each expression.

10. $(x+4)^6$

11. $(x-2)^3$

12. $(1-x^2)^6$

13. $(x+y)^6$

14. $(x-2y)^4$

15. $(x+3y)^8$

16. $(3+2x)^4$

17. $(3x-1)^5$

Write the indicated term of each binomial expansion.

18. seventh term of $(2x+y)^8$

19. fifth term of $(x-y)^{10}$

20. tenth term of $(x^2-1)^{11}$

Assume that n is a positive integer. Use the method of mathematical induction to prove each statement.

1. $1^2 + 3^2 + 5^2 + ... + (2n-1)^2 = \frac{1}{3}n(2n-1)(2n+1)$

2. $4 + 8 + 12 + ... + 4n = 2n(n+1)$

3. $5 + 7 + 9 + ... + (2n+3) = n(n+4)$

4. $1 \cdot 2 + 2 \cdot 3 + 3 \cdot 4 + ... + n(n+1) = \frac{1}{3}n(n+1)(n+2)$

5. $1 + 5 + 9 + ... + (4n-3) = n(2n-1)$

6. $5 + 8 + 11 + ... + (3n+2) = \frac{1}{2}n(3n+7)$

7. $1 \cdot 3 + 2 \cdot 4 + 3 \cdot 5 + ... + n(n+2) = \frac{1}{6}n(n+1)(2n+7)$

8. $2 + 5 + 8 + ... + (3n-1) = \frac{1}{2}n(3n+1)$

9. $2 + 2^2 + 2^3 + ... + 2^n = 2(2^n - 1)$

10. $\cos(n\pi) = (-1)^n$

11. $\frac{1}{1 \cdot 2 \cdot 3} + \frac{1}{2 \cdot 3 \cdot 4} + \frac{1}{3 \cdot 4 \cdot 5} + ... + \frac{1}{n(n+1)(n+2)} = \frac{n(n+3)}{4(n+1)(n+2)}$

12. $2^{n+3} < (n+3)!$

13. If $h \geq 0$, then $1 + nh \leq (1+h)^n$.

14. $n < 2^n$

15. $2 \leq 2^n$

16. If $n \geq 7$, then $\left(\frac{4}{3}\right)^n > n$.

17. If $n \geq 2$, then $\frac{1}{\sqrt{1}} + \frac{1}{\sqrt{2}} + \frac{1}{\sqrt{3}} + ... + \frac{1}{\sqrt{n}} > \sqrt{n}$.

18. If $n \geq 1$ and $0 < a < b$, then $\left(\frac{a}{b}\right)^{n+1} < \left(\frac{a}{b}\right)^n$.

19. $\left(\frac{a}{b}\right)^n = \frac{a^n}{b^n}$

11.6 Counting Theory

Evaluate each expression.

1. $C(6,4)$
2. $P(4,2)$
3. $C(5,3)$
4. $P(8,3)$

5. $_{10}P_7$
6. $_{20}C_4$
7. $_{28}P_2$
8. $_{15}C_6$

Use the fundamental principle of counting or permutations to solve each problem.

9. In how many ways can 2 cards be drawn in succession from a pack of 5 different cards marked !, @, #, $, and % ? Assume that the first card is not replaced before the second card is drawn.

10. In how many ways can 2 cards be drawn in succession from a pack of 5 different cards marked !, @, #, $, and % ? Assume that the first card is replaced before the second card is drawn.

11. A deli offers a value meal that consists of a sandwich, a bag of chips and a soft drink. Suppose there are 9 different kinds of sandwiches, 4 different kinds of chips, and 5 different kinds of soft drinks. How many different meals can be made?

12. In how many ways can 6 students be seated in a row of 6 seats?

13. How many different 2 digit numbers can be formed using the digits 2, 4, 6, and 8? Assume that the digits cannot be repeated.

14. How many different 3 digit numbers can be formed using the digits 2, 4, 6, and 8? Assume that the digits can be repeated.

15. In a club with 12 members, how many ways can a slate of 3 officers consisting of president, vice-president, and treasurer be chosen?

16. In how many ways can 9 players be assigned to the 4 positions on a team, assuming that any player can play any position?

17. Two nine sided dice, one pink and one white, are thrown. How many different outcomes are possible?

18. In a club with 9 women and 7 men members, how many 5-member committees can be chosen that have all men?

19. In a club with 9 women and 7 men members, how many 5-member committees can be chosen that have 4 women and 3 men?

11.7 Basics of Probability.

1. Find the probability of picking either an ace or a king in a single drawing from a standard deck of cards.

2. An urn contains 3 red balls, 4 blue balls, 2 green balls, and 1 yellow ball. What is the probability of drawing a blue ball?

3. An urn contains 3 red balls, 4 blue balls, 2 green balls, and 1 yellow ball. If two balls are drawn in succession what is the probability of drawing a blue ball followed by a yellow ball?

4. Find the probability of rolling a 5 on a single roll of a die.

5. Find the probability of rolling a sum of 6 when 2 dice are rolled.

6. Find the probability of picking a red card from a deck of cards.

7. Find the probability of rolling a sum greater than 4 when 2 dice are rolled.

8. The probability of being left-handed is about .10. What are the odds against being left-handed?

9. What is the probability of throwing a six twice in succession with a single die?

10. Find the probability of obtaining no heads in tossing a pair of coins.

11. An urn contains 3 red balls, 4 blue balls, 2 green balls, and 1 yellow ball. What are the odds in favor of drawing a green ball?

12. An urn contains 3 red balls, 4 blue balls, 2 green balls, and 1 yellow ball. What are the odds against drawing a red ball?

13. An urn contains 3 red balls, 4 blue balls, 2 green balls, and 1 yellow ball. What is the probability of drawing 2 yellow balls in succession without replacing the first ball drawn?

14. What is the probability of picking an "i" from the word "Mississippi"?

15. A multiple-choice test question has five choices a – e. What is the probability of guessing the correct answer?

16. The digits 0 – 9 are written on a slip of paper and placed in a hat. What is the probability of selecting an odd digit if one slip is drawn from the hat?

17. The digits 0 – 9 are written on a slip of paper and placed in a hat. What is the probability of selecting a digit greater than 4 or an even digit if one slip is drawn from the hat?

11.1 Sequences and Series

1. 2, 6, 12, 20, 30
2. $-\dfrac{2}{3}, \dfrac{1}{3}, 2, \dfrac{13}{3}, \dfrac{22}{3}$
3. −1, 1, −1, 1, −1
4. 1, 3, 7, 15, 31
5. 4, 7, 10, 13,
6. 5, 2, −1, −4
7. 2, 4, 12, 48
8. 81, 27, 9, 3
9. 55
10. 204
11. $\dfrac{71}{20}$
12. 255
13. 35
14. $\dfrac{251}{216}$
15. 3999.27
16. 6141
17. $\dfrac{2}{5}$
18. 3
19. 30
20. 330
21. 165
22. 515

11.2 Arithmetic Sequences and Series

1. −2
2. 1.5
3. 0.25
4. $x+2$
5. 7, 11, 15, 19, 23
6. $9, \dfrac{28}{3}, \dfrac{29}{3}, 10, \dfrac{31}{3}$
7. $a_n = 50 + 15n;\ a_5 = 125$
8. $a_n = -11 + 5n;\ a_5 = 14$
9. $a_n = 2n;\ a_5 = 10$
10. $a_n = -15 + 5n;\ a_5 = 10$
11. −32
12. 30
13. 80
14. −52
15. $a_1 = \dfrac{45}{7};\ d = -\dfrac{17}{7}$
16. $a_1;\ d = -20 - a_1$
17. 20
18. 130
19. 64
20. 93

11.3 Geometric Sequences and Series

1. $a_n = 8\left(\dfrac{1}{2}\right)^{n-1}$; $a_4 = 1$

2. $a_n = -2(-3)^{n-1}$; $a_4 = 54$

3. $a_n = 4^{n-1}$; $a_4 = 64$

4. $a_n = 9\left(\dfrac{1}{3}\right)^{n-1}$; $a_4 = \dfrac{1}{3}$

5. $a_1 = 27$; $r = \dfrac{1}{3}$

6. $a_1 = -24$; $r = \dfrac{1}{2}$

7. $\dfrac{1705}{64}$

8. $\dfrac{62}{5}$

9. 1562

10. 61

11. $\dfrac{781}{3125}$

12. $-\dfrac{237}{1000}$

13. 726

14. $\dfrac{555}{64}$

15. $r = -3$, does not converge

16. $r = \dfrac{1}{5}$

17. $\dfrac{1}{2}$

18. $\dfrac{100}{9}$

19. does not converge

20. 3

11.4 The Binomial Theorem

1. 126
2. 10
3. 15
4. 20
5. 4

6. 21
7. 1
8. 35
9. 15

10. $x^6 + 24x^5 + 240x^4 + 1280x^3 + 3840x^2 + 6144x + 4096$

11. $x^3 - 6x^2 + 12x - 8$

12. $x^{12} - 6x^{10} + 15x^8 - 20x^6 + 15x^4 - 6x^2 + 1$

13. $x^6 + 6x^5y + 15x^4y^2 + 20x^3y^3 + 15x^2y^4 + 6xy^5 + y^6$

14. $x^4 - 8x^3y + 24x^2y^2 - 32xy^3 + 16y^4$

15. $x^8 + 24x^7y + 252x^6y^2 + 1512x^5y^3 + 5670x^4y^4 + 13608x^3y^5 + 20412x^2y^6 + 17496xy^7 + 6561y^8$

16. $16x^4 + 96x^3 + 216x^2 + 216x + 81$

17. $243x^5 - 405x^4 + 270x^3 - 90x^2 + 15x - 1$

18. $112x^2 y^6$

19. $210x^6 y^4$

20. $-55x^4$

11.6 Counting Theory

1. 15   **2.** 12   **3.** 10   **4.** 336

5. 604,800   **6.** 4845   **7.** 756   **8.** 5005

9. 20   **10.** 25   **11.** 180   **12.** 720

13. 12   **14.** 64   **15.** 1320   **16.** 126

17. 81   **18.** 21   **19.** 4410

11.7 Basics of Probability

1. $\frac{2}{13}$ **2.** $\frac{2}{5}$ **3.** $\frac{2}{45}$ **4.** $\frac{1}{6}$

5. $\frac{5}{36}$ **6.** $\frac{1}{2}$ **7.** $\frac{5}{6}$ **8.** $9:1$

9. $\frac{1}{36}$ **10.** $\frac{1}{4}$ **11.** $1:4$ **12.** $7:3$

13. 0   **14.** $\frac{4}{11}$ **15.** $\frac{1}{5}$ **16.** $\frac{1}{2}$

17. $\frac{4}{5}$

Identify the type of conic represented.

1. $r = \dfrac{2}{1-\cos\theta}$

2. $r = \dfrac{2}{2-\cos\theta}$

3. $r = \dfrac{3}{2+4\sin\theta}$

4. $r = \dfrac{3}{4-2\cos\theta}$

5. $r = \dfrac{5}{3+3\cos\theta}$

6. $r = \dfrac{2}{1+\sqrt{2}\cos\theta}$

7. $r = \dfrac{8}{2-3\sin\theta}$

8. $r = \dfrac{4}{2+\cos\theta}$

9. $r = \dfrac{12}{3-6\cos\theta}$

10. $r = \dfrac{9}{3-3\sin\theta}$

Find a polar equation for the conic with focus at the pole, satisfying the given conditions.

11. $e = 1$, vertical directrix 1 unit to the left of the pole

12. $e = \dfrac{1}{2}$, horizontal directrix 1 unit below the pole

13. $e = 2$, vertical directrix 1 unit to the right of the pole

14. $e = \dfrac{2}{3}$, vertical directrix 5 units to the right of the pole

15. $e = 3$, horizontal directrix 2 units below the pole

16. $e = 2$, vertical directrix 1 unit to the left of the pole

17. $e = 1$, horizontal directrix 2 units above the pole

18. $e = \dfrac{2}{3}$, horizontal directrix 4 units below the pole

19. $e = 1$, horizontal directrix 2 units below the pole

20. $e = \dfrac{3}{4}$, horizontal directrix 2 units below the pole

128   Appendix B Rotation of Axes

Find the angle of rotation θ that will remove the *xy*-term in each equation.

1. $7x^2 - 6\sqrt{3}xy + 13y^2 - 16 = 0$

2. $3x^2 + 2xy + 3y^2 - 32y + 92 = 0$

3. $9x^2 - 24xy + 16y^2 - 320x - 240y = 0$

4. $5x^2 - 6\sqrt{3}xy - 11y^2 + 4x - 3y + 2 = 0$

5. $3x^2 - \sqrt{3}xy + 2y^2 - 3x + y - 10 = 0$

6. $5x^2 - 8xy + 5y^2 + y + 3 = 0$

7. $6x^2 + \sqrt{3}xy + 7y^2 - 36 = 0$

8. $x^2 + xy + y^2 - 1 = 0$

9. $3x^2 + 4\sqrt{3}xy - y^2 - 7 = 0$

10. $x^2 - 6xy - 5y^2 - 27 = 0$

In each equation, remove the *xy*-term by rotation.

11. $6x^2 - 6xy + 14y^2 - 45 = 0$

12. $x^2 + 4xy - 2y^2 - 1 = 0$

13. $5x^2 + 26xy + 5y^2 - 72 = 0$

14. $x^2 - 2\sqrt{3}xy - y^2 - 2 = 0$

15. $6x^2 - 24xy - y^2 - 30 = 0$

16. $5x^2 - 4xy + 8y^2 - 36 = 0$

17. $13x^2 + 6\sqrt{3}xy + 7y^2 - 16 = 0$

18. $17x^2 + 32xy - 7y^2 - 75 = 0$

19. $5x^2 - 2xy + 5y^2 - 12 = 0$

20. $2x^2 - 3xy - 2y^2 + 10 = 0$

Appendix A: Polar Form of Conic Sections

1. parabola
2. ellipse
3. hyperbola
4. ellipse
5. parabola
6. hyperbola
7. hyperbola
8. ellipse
9. hyperbola
10. parabola
11. $r = \dfrac{1}{1 - \cos\theta}$
12. $r = \dfrac{1}{2 + \sin\theta}$
13. $r = \dfrac{2}{1 + 2\cos\theta}$
14. $r = \dfrac{10}{3 + 2\cos\theta}$
15. $r = \dfrac{6}{1 - 3\sin\theta}$
16. $r = \dfrac{2}{1 - 2\cos\theta}$
17. $r = \dfrac{2}{1 + \sin\theta}$
18. $r = \dfrac{8}{3 - 2\sin\theta}$
19. $r = \dfrac{2}{1 - \sin\theta}$
20. $r = \dfrac{6}{4 - 3\sin\theta}$

Appendix B: Rotation of Axes

1. $30°$
2. $45°$
3. $36.9°$
4. $73.5°$
5. $60°$
6. $45°$
7. $60°$
8. $45°$
9. $30°$
10. $67.5°$
11. $\dfrac{x'^2}{9} + \dfrac{y'^2}{3} = 1$
12. $\dfrac{x'^2}{5/8} - \dfrac{y'^2}{1/3} = 1$
13. $\dfrac{x'^2}{4} - \dfrac{y'^2}{9} = 1$
14. $y'^2 - x'^2 = 1$
15. $\dfrac{x'^2}{2} - \dfrac{y'^2}{3} = 1$
16. $\dfrac{x'^2}{9} + \dfrac{y'^2}{4} = 1$
17. $x'^2 + \dfrac{y'^2}{4} = 1$
18. $\dfrac{x'^2}{3} - \dfrac{y'^2}{5} = 1$
19. $\dfrac{x'^2}{3} + \dfrac{y'^2}{2} = 1$
20. $\dfrac{x'^2}{4} - \dfrac{y'^2}{4} = 1$